PHILIP'S

BRITAIN & IRELAND

NIGEL HENBEST

STARGAZING 2027

MONTH-BY-MONTH GUIDE TO THE NIGHT SKY

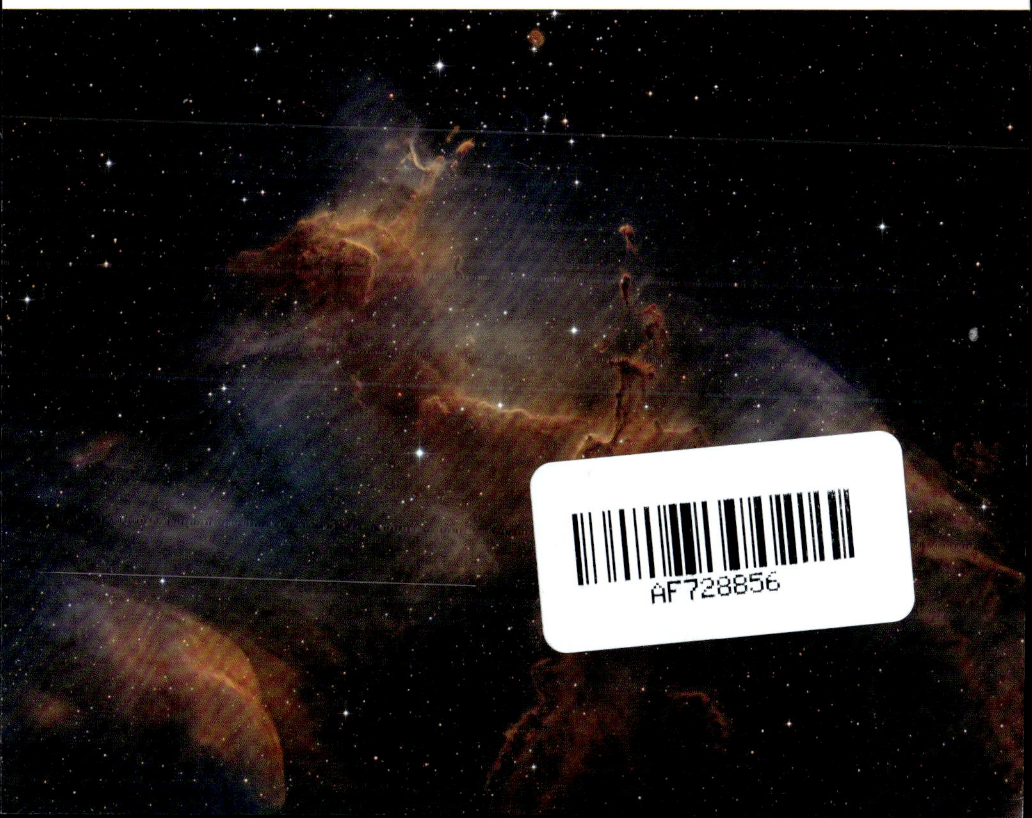

www.philips-maps.co.uk

First published in Great Britain in 2026
by Philip's, an imprint of
Octopus Publishing Group Ltd
Carmelite House
50 Victoria Embankment
London EC4Y 0DZ
www.octopusbooks.co.uk

An Hachette UK Company
www.hachette.co.uk

The authorized representative in the EEA is Hachette
Ireland, 8 Castlecourt Centre, Dublin 15, D15 XTP3, Ireland
(email: info@hbgi.ie)

ISBN 978-1-84907-755-2
eISBN 978-184907-756-9

A CIP catalogue record for this book is available from the
British Library.

Printed in China
10 9 8 7 6 5 4 3 2 1

Cover: Dorset coast sky with Orion.
Title page: Gum Nebula.

CONTENTS

Welcome to the latest edition of *Stargazing*! Within these pages, you'll find a complete guide to everything happening in the night sky throughout 2027 – whether you're a beginner or an experienced astronomer.

With the 12 monthly Star Charts, you can find your way around the sky on any night in the year. Impress your friends by identifying celestial sights ranging from the brightest planets to some pretty obscure constellations.

Every page of *Stargazing 2027* is bang up-to-date, bringing you everything that's new this year, from shooting stars to eclipses. And I'll start with a run-down of the most exciting celestial events on view in 2027 (opposite).

THE STAR CHARTS

A reliable map is just as essential for exploring the heavens as it is for visiting a foreign country. So each monthly section starts with a circular **Star Chart** showing the whole evening sky.

To keep the maps uncluttered, I've plotted about 200 of the brighter stars (down to third magnitude), which means you can pick out the main star patterns – the constellations. (If the charts showed every star visible on a really dark night, there'd be around 3000 stars on each!) I also show the ecliptic: the apparent path of the Sun in the sky; it's closely followed by the Moon and planets as well.

You can use these charts throughout the UK and Ireland, along with most of Europe, North America and northern Asia – between 40 and 60 degrees north – though the precise timings apply specifically to Britain and Ireland.

USING THE STAR CHARTS

It's pretty easy to use the charts. Start by working out your compass points. South is where the Sun is highest in the sky during the day; east is roughly where the Sun rises, and west where it sets. At night, you can find north by locating the Pole Star – Polaris – by using the stars of the Plough (see May's Constellation).

The left-hand chart then shows your view to the north. Most of the stars here are visible all year: these circumpolar constellations wheel around Polaris as the seasons progress. Your view to the south appears in the right-hand chart; it changes much more as the Earth orbits the Sun. Leo's prominent 'Sickle' is high in the spring skies. Summer is dominated by bright Vega, Deneb and Altair. Autumn's familiar marker is the Square of Pegasus; while the stars of Orion rule the winter sky.

During the night, our perspective on the sky also alters as the Earth spins round, making the stars and planets appear to rise in the east and set in the west. The charts depict the sky in the late evening (the exact times are noted in the captions). As a rule of thumb, if you are observing two hours later, then the following month's chart will be a better guide to the stars on view – though beware: the Moon and planets won't be in the right place.

THE PLANETS, MOON AND SPECIAL EVENTS

The charts also highlight the **planets** above the horizon in the late evening. I've indicated the track of any **comets** known at the time of writing; though I can't guide you to a comet that's found after the book has been printed!

HIGHLIGHTS OF THE YEAR

- **2–4 January, before dawn:** the crescent Moon passes near Venus.
- **3 January:** Venus reaches its greatest morning separation from the Sun.
- **Night of 3/4 January:** an excellent year for catching the Quadrantid meteor shower.
- **22 January:** the year's biggest and brightest supermoon passes in front of Praesepe.
- **23 January:** the Moon is very near Jupiter.
- **24 January:** the Moon lies near to Mars.
- **1–3 February, before dawn:** the crescent Moon passes Venus with Antares nearby.
- **6 February:** an annular eclipse of the Sun is visible from South America and western Africa.
- **10 February:** Jupiter closest to the Earth.
- **11 February:** Jupiter lies opposite the Sun.
- **19 February:** Mars lies opposite the Sun; Moon near Jupiter.
- **20 February:** Mars closest to the Earth and near the Full Moon.
- **4–5 March,** before dawn: Venus lies near the crescent Moon.
- **Night of 13/14 March:** the Moon occults the Pleiades.
- **19 March:** the Moon forms a tight group with Jupiter, Mars and Regulus.
- **15 April:** the Moon is near Jupiter and Mars.
- **Night of 22/23 April:** the Lyrid meteor shower is spoilt by bright moonlight.
- **7 May, early hours:** an excellent year for observing the Eta Aquarid meteor shower.
- **12–13 May:** the Moon is near to Jupiter, Mars and Regulus.
- **14 May:** Mars passes just above Regulus.
- **28 May:** Mercury is at its greatest separation from the Sun, in its best evening appearance of the year.
- **8–10 June:** the Moon passes Jupiter, Regulus and Mars.
- **6 July:** crescent Moon near Jupiter.
- **2 August:** a total solar eclipse is visible from a narrow band of north Africa. The UK and Ireland experience a partial eclipse (20 to 45 per cent depending on your location).
- **Night of 12/13 August:** the Perseid meteor shower is spoilt by bright moonlight.
- **Night of 24/25 August:** the Moon occults the Pleiades.
- **21 September, am:** the Moon occults the Pleiades.
- **27 September:** Neptune is closest to the Earth.
- **28 September, am:** the crescent Moon lies near Jupiter.
- **28 September:** Neptune is opposite the Sun.
- **15 October:** the Full Moon above Saturn.
- **18 October:** Saturn lies opposite the Sun and closest to the Earth.
- **Night of 21/22 October:** the Orionid meteor shower is spoilt by moonlight.
- **26 October, am:** the Moon near Jupiter.
- **1 November:** the crescent Moon lies near Venus with Mars between them.
- **4 November:** Mercury reaches its greatest separation from the Sun in its best morning appearance of the year.
- **14 November:** the Full Moon occults the Pleiades.
- **Night of 17/18 November:** moonlight washes out the Leonid meteor shower.
- **24–25 November:** Mars very close to Venus.
- **30 November:** the crescent Moon lies next to Venus, with Mars to the right.
- **30 November:** Uranus is opposite the Sun and nearest to the Earth.
- **12 December, am:** the Moon occults the Pleiades.
- **Nights of 13/14 and 14/15 December:** the Geminid meteor shower is spoilt by bright moonlight.
- **30 December:** Venus near the crescent Moon.

I've plotted the position of the Full Moon each month, and also the **Moon's position** at three-day intervals before and afterwards. If there's a **meteor shower** in the month, the charts show its radiant – the position from which the meteors stream outwards.

The **Calendar** provides a daily guide to the Moon's phases and other celestial happenings. I've detailed the most interesting in the **Special Events** section, including close pairings of the planets, times of the equinoxes and solstices and – most exciting – **eclipses** of the Moon and Sun.

Check out the **Planet Watch** page for more about the other worlds of the Solar System, including their antics at times they're not on the monthly Star Charts. I've illustrated unusual planetary and lunar goings-on in the **Special Event Charts**. There's a full guide to planetary motions, eclipses and meteor showers in **Solar System 2027** on pages 80–82.

MONTHLY OBJECTS, TOPICS AND PICTURES

Each monthly section highlights a fascinating **object** (such as a planet, a star or a nebula) and a stunning **picture** taken by a British or Irish amateur astronomer – along with full technical information. There's also an in-depth exploration of an intriguing and often newsworthy **topic**, ranging from Polynesian astronomy to black holes, and from peculiar star names to your chance to peek at the far side of the Moon.

GETTING IN DEEP

A practical **Observing Tip** is included each month, helping you to explore the sky with the naked eye, binoculars or a telescope.

Check out my guide to the **Top 20 Sky Sights**, including nebulae, star clusters and galaxies. You'll find it on pages 83–85.

Equipment expert Robin Scagell provides his insights into the best – and safest – ways to observe the Sun and to take stunning close-up images of sunspots and other details of its ever-changing magnetic weather, on pages 86–89.

The final section details the internationally approved dark-sky sites in Britain and Ireland, where you're guaranteed to be free of light pollution. It also includes a round-up of planetariums and public observatories you can visit, plus a guide to the best star camps and astronomical festivals.

Happy stargazing!

JARGON BUSTER

Have you ever wondered how astronomers describe the brightness of the stars or how far apart they appear in the sky? Not to mention how we can measure the distances to the stars? If so, you can quickly find yourself mired in some arcane astro-speak – magnitudes, arcminutes, light years and the like.

Here's a quick and easy guide to busting the astronomical jargon.

Magnitudes

It only takes a glance at the sky to see that some stars are pretty brilliant, while many more are dim. But how do we describe to other people how bright a star appears?

Around 2000 years ago, ancient Greek astronomers ranked the stars into six classes, or **magnitudes**, depending on their brightness. The most brilliant stars were first magnitude, and the faintest stars you can see came in at sixth magnitude. So the stars of the Plough, for instance, are second magnitude while the individual Seven Sisters in the Pleiades are fourth magnitude.

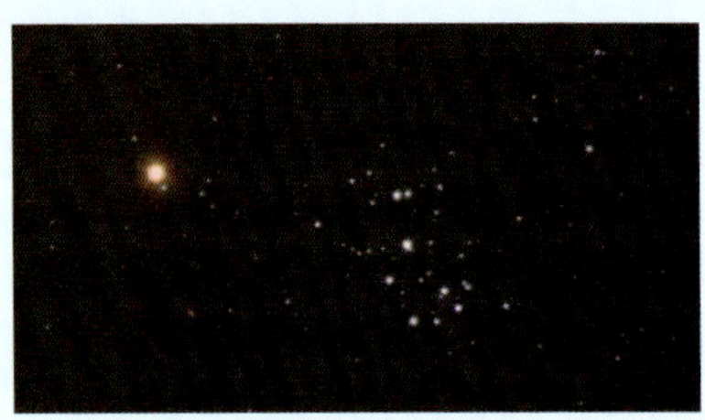

Mars (magnitude –1.6 here) shines a hundred times brighter than the Seven Sisters in the Pleiades, which are around five magnitudes fainter.

Today, scientists can measure the light from the stars with amazing accuracy. (Mathematically speaking, a difference of five magnitudes represents a difference in brightness of one hundred times.) So the Pole Star is magnitude +2.0, while Rigel is magnitude +0.1. Because we've inherited the ancient ranking system, the brightest stars have the *smallest* magnitude. In fact, the most brilliant stars come in with a negative magnitude, including Sirius (magnitude −1.5).

And we can use the magnitude system to describe the brightness of other objects in the sky, such as stunning Venus, which can be almost as brilliant as magnitude −5. The Full Moon and the Sun have whopping negative magnitudes!

At the other end of the scale, stars, nebulae and galaxies with a magnitude fainter than +6.5 are too dim to be seen by the naked eye. Using ever larger telescopes – or by observing from above Earth's atmosphere – you can perceive fainter and fainter objects. The most distant galaxies visible to the Hubble Space Telescope are ten billion times fainter than the naked-eye limit.

Here's a guide to the magnitude of some interesting objects:

Sun	−26.7
Full Moon	−12.5
Venus (at its brightest)	−4.9
Sirius	−1.5
Betelgeuse (variable)	0.0 – +1.6
Polaris (Pole Star)	+2.0
Faintest star visible to the naked eye	+6.5
Faintest star visible to the Hubble Space Telescope	+31

Degrees of separation

Astronomers measure the distance between objects in the sky in **degrees** (symbol °): all around the horizon is 360°, while it's 90° from the horizon to the point directly overhead (the zenith).

As you can see in the photograph, it's possible to use your hand – held at arm's length – to give a rough idea of angular distances in the sky.

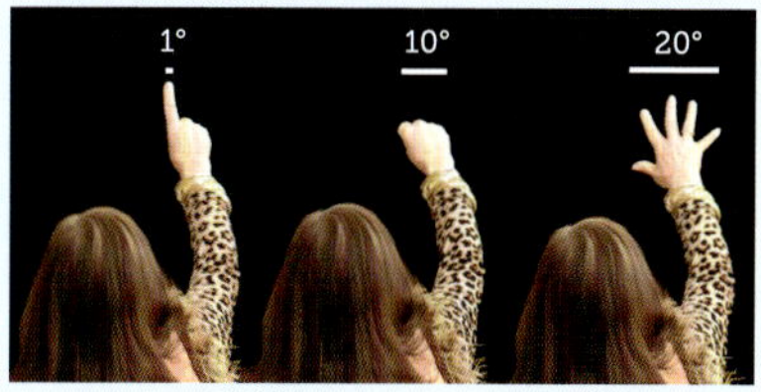

For objects that are very close together – like many double stars – we divide the degree into 60 arcminutes (symbol ′). And for celestial objects that are extremely tiny – such as the discs of the planets – we split each arcminute into 60 arcseconds (symbol ″). To give you an idea of how small these units are, it takes 3600 arcseconds to make up one degree.

Here are some typical separations and sizes in the sky:

Length of the Plough	25°
Width of Orion's Belt	3°
Diameter of the Moon	31′
Separation of Mizar and Alcor	12′
Diameter of Jupiter	45″
Separation of Albireo A and B	35″

How far's that star?

Everything we see in the heavens lies a long way off. We can give distances to the planets in millions of kilometres. But the stars are so distant that even the nearest, Proxima Centauri, lies some 40 million million kilometres away. To turn those distances into something more manageable, astronomers use a larger unit: one **light year** is the distance that light travels in a year.

One light year is about 9.46 million million kilometres. That makes Proxima Centauri a much more manageable 4.2 light years away from us. Here are the distances to some other familiar astronomical objects, in light years:

Sirius	8.6
Polaris	440
Centre of the Milky Way	26,700
Andromeda Galaxy	2.5 million
Most distant galaxies seen by James Webb Space Telescope	13.5 billion

- The sky at 10 pm in mid-January, with Moon positions at three-day intervals either side of Full Moon.
- The star positions are also correct for 11 pm at the beginning of January, and 9 pm at the end of the month.
- The planets move slightly relative to the stars during the month.

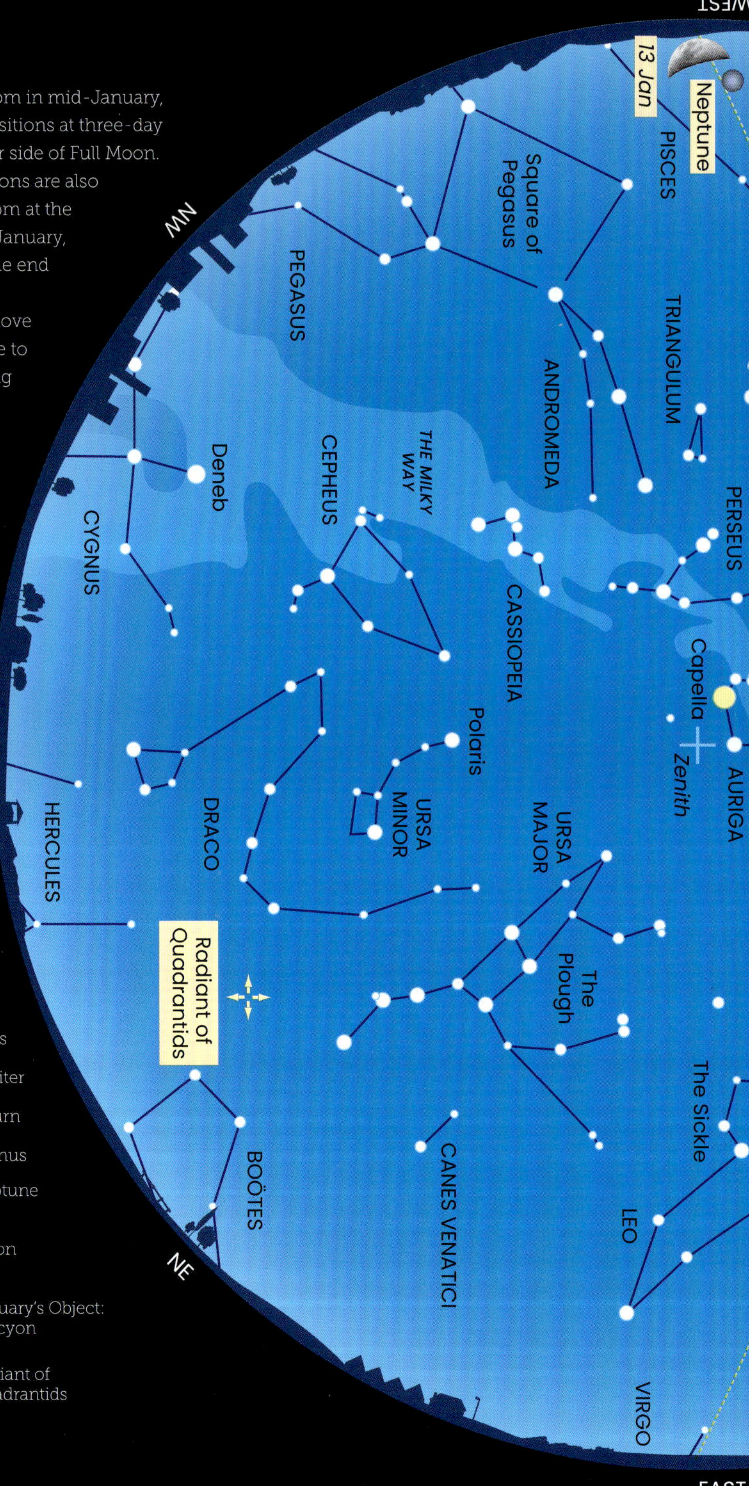

- Mars
- Jupiter
- Saturn
- Uranus
- Neptune
- Moon
- January's Object: Procyon
- Radiant of Quadrantids

JANUARY

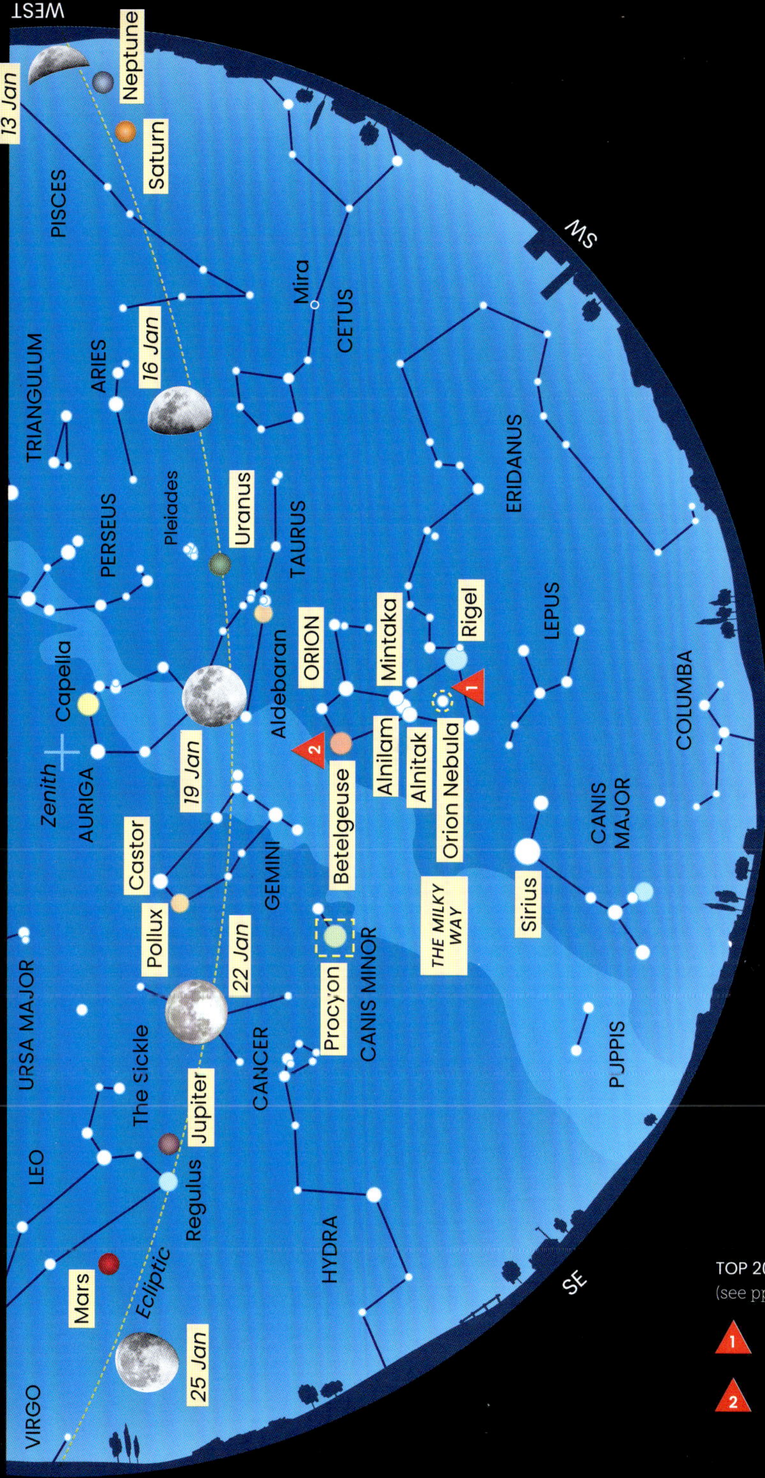

TOP 20 SKY SIGHTS
(see pp. 83–85)

1 Orion Nebula

2 Betelgeuse

The evening sky boasts a trio of bright planets – **Saturn**, **Jupiter** and **Mars** – while Venus is glorious before dawn. Throw in the year's best display of shooting stars, a supermoon and the dazzling winter constellations, and the new year is off to a spectacular start.

JANUARY'S CONSTELLATION

The scintillating stellar figure of **Orion** strides high across the heavens this month. According to various ancient cultures, this constellation represented a shepherd, a mighty king or even a canoe; but to the ancient Greeks he was a formidable hunter, and also the world's most handsome man. In the sky Orion also cuts a dash, containing one-tenth of the brightest stars.

Blood-red **Betelgeuse** marks one of his shoulders. This red giant is 760 times wider than the Sun, and one day it's destined to blow itself apart as a supernova. Orion's other brilliant star, **Rigel**, lies at the bottom of the hunter's tunic. In contrast to Betelgeuse, it's blue-white in colour, with a searingly hot temperature of 12,000°C.

Orion's Belt is marked by well-matched **Alnitak**, **Alnilam** and **Mintaka**, lying over 1200 light years away. Despite this immense distance, they're prominent in our skies because each outshines our Sun 200,000 times over.

Look carefully below the Belt to spot a faint patch of light representing Orion's sword. This is the great **Orion Nebula**, a seething maelstrom of incandescent gas that's the birthplace of new stars created from a dark cloud of dust and gas.

OBSERVING TIP

Hold a meteor party to check out the Quadrantid meteor shower on the night of 3/4 January. You don't need any optical equipment: the ideal viewing equipment is your unaided eye, plus a warm sleeping bag and a lounger. Everyone should look in different directions, so you can cover the whole sky. Shout out 'meteor!' when you see a shooting star. One of your party can record the observations, using a watch, notepad and red torch.

JANUARY'S OBJECT

It's the eighth brightest star in the night sky, but **Procyon** always seems to play Cinderella to the headline-grabbing stars around it: curiously named Betelgeuse, the celestial twins **Castor** and **Pollux**, and brilliant **Sirius**, the Dog Star.

The name Procyon is Greek for 'before the dog', because Mediterranean people saw this star rising shortly before Sirius. In Arabic folklore, these two were sisters: Sirius crossed the great sky-river (the Milky Way), leaving Procyon behind. To the Polynesians, Sirius was the body of a vast seabird, with Procyon and Canopus (below the horizon from the UK) forming its wingtips.

In reality, Procyon is a fairly ordinary star, 50 per cent larger than the Sun and seven times more luminous. It's bright in our skies only because it lies just 11.5 light years away. And it's a star with a secret: a shrunken white dwarf companion, Procyon B, as heavy as 0.6 Suns but only a little larger than the Earth. By a strange coincidence, the only other first-magnitude star with a white dwarf in attendance is its sister star, Sirius.

JANUARY'S TOPIC: ENCKE'S COMET

Grab your binoculars this month for a grandstand view of Encke's Comet – only the second comet to have its orbit determined (after Halley's Comet), by German mathematician Johann Franz Encke (1791–1865).

Encke's Comet orbits the Sun every 3.3 years, and it has been observed on 65 close passes of the Sun. Always a dim object, it was last easily visible in 2003. The comet reaches magnitude +5 in late January as it heads towards its next solar encounter on 10 February. Binoculars will show you a fuzzy ball of light in Aquarius (see Special Events).

At its centre lies a solid nucleus of ice and rock 5 kilometres across. Astronomers have discovered 88 other icy and rocky fragments following the same orbit – all probably debris from a single disintegrated mega-comet. One of these chunks was responsible for widespread devastation when it hit the Tunguska region of Siberia in 1908.

JANUARY'S PICTURE

Venus starts the year as a brilliant Morning Star, reaching its greatest separation from the Sun on 3 January, and through a telescope you'll see it as a globe half-lit by the Sun. But even the best telescope won't reveal its surface: the entire planet is cloaked in clouds of sulphuric acid droplets, forming – to the human eye – an all-enveloping and featureless pall.

But turn to astrophotography, as Tom Williams has done, expand your wavelength range to include ultraviolet and infrared, and a whole new world is on display. Tom took advantage of excellent conditions in Earth's atmosphere on 1 June 2023 to capture these fine details in Venus's atmosphere. Near the equator, the clouds swirl in giant whorls as they're swept around the planet by 4000 kilometre-per-hour winds, rising and falling over the mountains below. At higher latitudes, the clouds move more smoothly as they form giant vortices around Venus's poles.

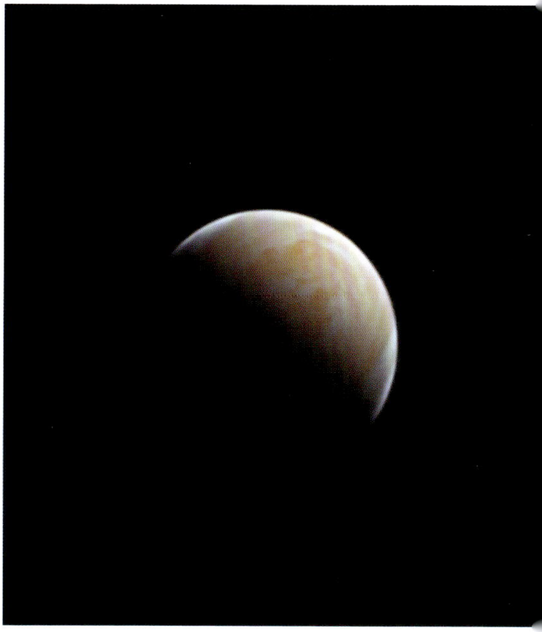

Tom Williams used a 400-mm Sky-Watcher 400P GoTo Dobsonian, operated at f/12 with a ZWO ASI462MM camera. He took 4 × 120-second exposures through an infrared filter, plus 6 × 120-second exposures through an ultraviolet filter, and created this image by coding the infrared image red, and the ultraviolet image blue (together with a green image that was the average of his infrared and ultraviolet views).

JANUARY'S CALENDAR

SUNDAY	MONDAY	TUESDAY	WEDNESDAY	THURSDAY	FRIDAY	SATURDAY
31 Moon near Antares (am)					1	2 Moon near Venus (am)
3 Earth at perihelion; Venus W elongation with Moon nearby (am); Quadrantids	4 Quadrantids (am); Moon near Venus (am)	5	6	7 8.24 pm New Moon	8	9
10	11	12	13	14 Moon near Saturn	15 8.34 pm First Quarter Moon	16
17	18	19	20	21	22 12.17 pm Full Moon; super-moon; Moon occults Praesepe	23 Moon very near Jupiter
24 Moon near Mars	25	26	27	28 Moon near Spica (am)	29 10.55 am Last Quarter Moon	30

SPECIAL EVENTS

- **2 January, before dawn:** a lovely sight as the crescent Moon lies to the right of Venus (Chart 1a).
- **3 January, 2.33 am:** the Earth is at perihelion, just 147 million km from the Sun.
- **3 January, before dawn:** Venus reaches its greatest morning separation from the Sun and forms a stunning duo with the Moon (Chart 1a).
- **Night of 3/4 January:** the maximum of the **Quadrantid meteor shower**, dust particles from the old comet 2003 EH_1 burning up in the Earth's atmosphere. With the Moon well out of the way, it's a great year for catching these bright colourful shooting stars.
- **4 January, before dawn:** you'll find the crescent Moon to the lower left of Venus (Chart 1a).
- **14 January:** the bright 'star' below the Moon is Saturn.
- **22 January, 7–9 pm:** the year's biggest and brightest supermoon (357,644 km away) passes in front of Praesepe.
- **23–24 January:** the Moon lies near Jupiter, Regulus and Mars (Chart 1b).
- **Last week of January:** Encke's Comet (magnitude +5) moves through Aquarius: best seen with binoculars low in the west around 6 pm, to the upper right of Mercury.

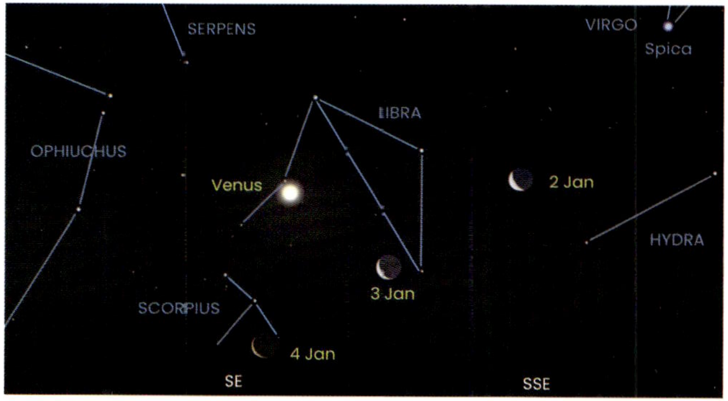

1a *2–4 January, 6 am. Venus and the crescent Moon.*

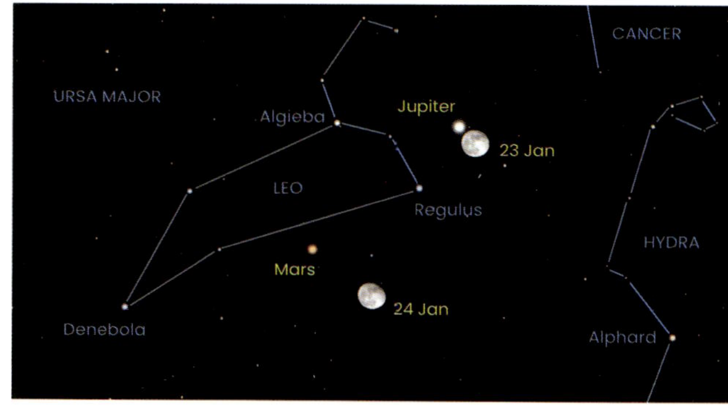

1b *23–24 January, 11 pm. The Moon with Jupiter, Mars and Regulus.*

- **Saturn**, over in the western sky, lies in Cetus, adjacent to the zodiacal constellation of Pisces. The ringworld shines at magnitude +0.8, and sets around 11 pm. The Moon is nearby on 14 January.
- Just below Saturn, binoculars or a telescope will show you **Neptune**, in Pisces. The distant dim (magnitude +7.8) planet sets about 10.30 pm.
- **Uranus** is in Taurus, near the Pleiades. Setting around 4 am, the seventh planet is just visible to the naked eye at magnitude +5.7.
- Rising about 7 pm, **Jupiter** lies in Leo, to the right of its brightest star, Regulus. The giant planet outshines all the stars, at magnitude −2.5. The Moon – just a day past Full – sails close by Jupiter on 23 January (Chart 1b).

Also in Leo, **Mars** lies to the lower left of Regulus, under the lion's belly. The Red Planet shines at magnitude −0.6 and rises around 8 pm. The Moon lies nearby on 24 January (Chart 1b).
- **Venus** – rising in the south-east about 4.30 am – blazes in the dawn sky at magnitude −4.5 and reaches its greatest separation from the Sun on 3 January. The Moon performs a picturesque pas de deux with the Morning Star on the mornings of 2, 3 and 4 January (Chart 1a).
- During the last few days of the month, **Mercury** appears low in the west after sunset at magnitude −0.9.

JANUARY'S PLANET WATCH

- The sky at 10 pm in mid-February, with Moon positions at three-day intervals either side of Full Moon.
- The star positions are also correct for 11 pm at the beginning of February, and 9 pm at the end of the month.
- The planets move slightly relative to the stars during the month.

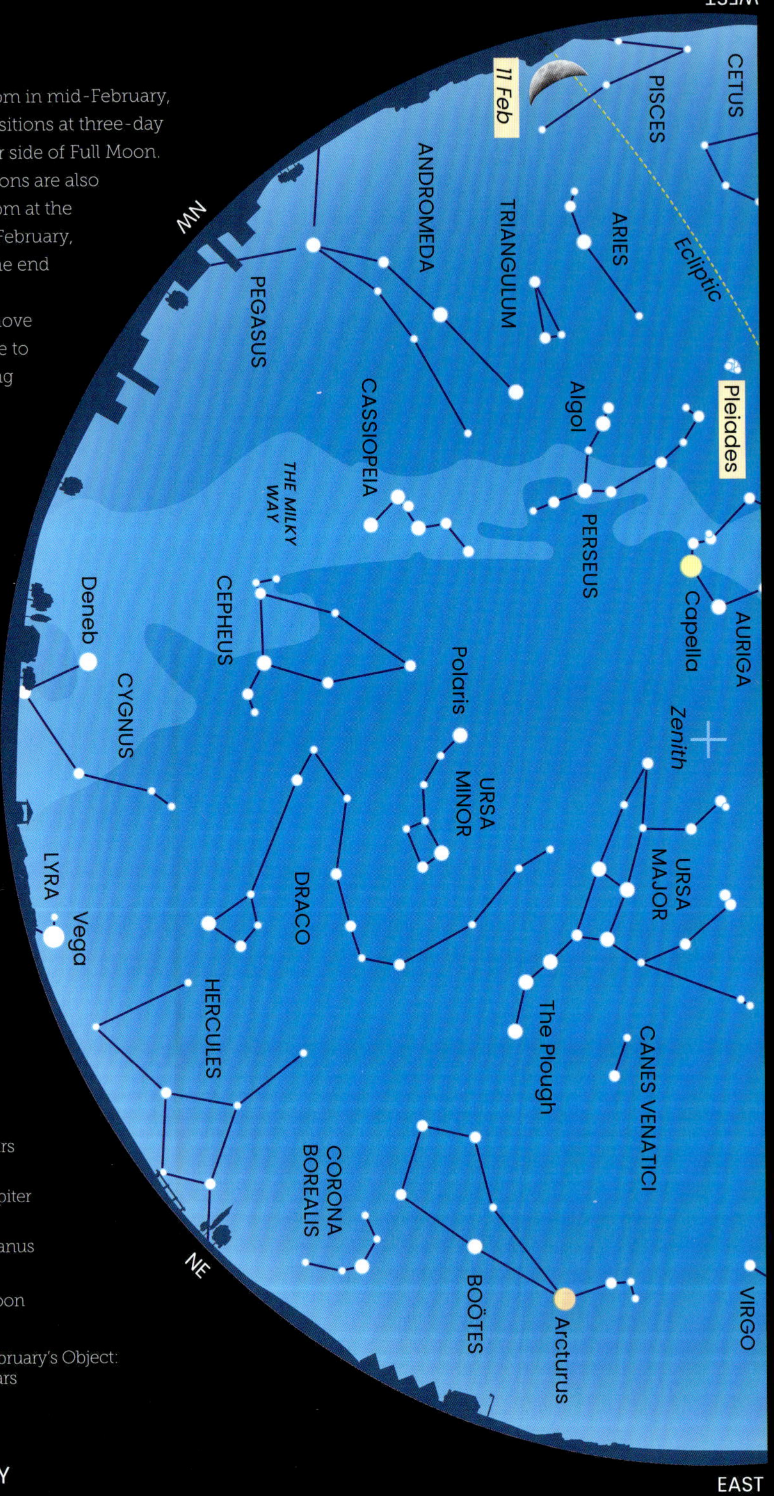

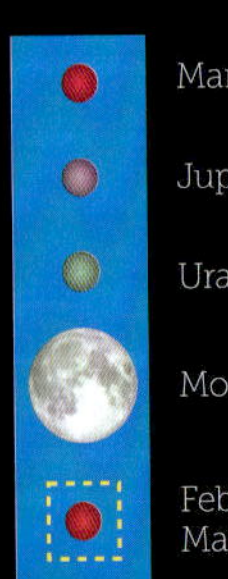

FEBRUARY

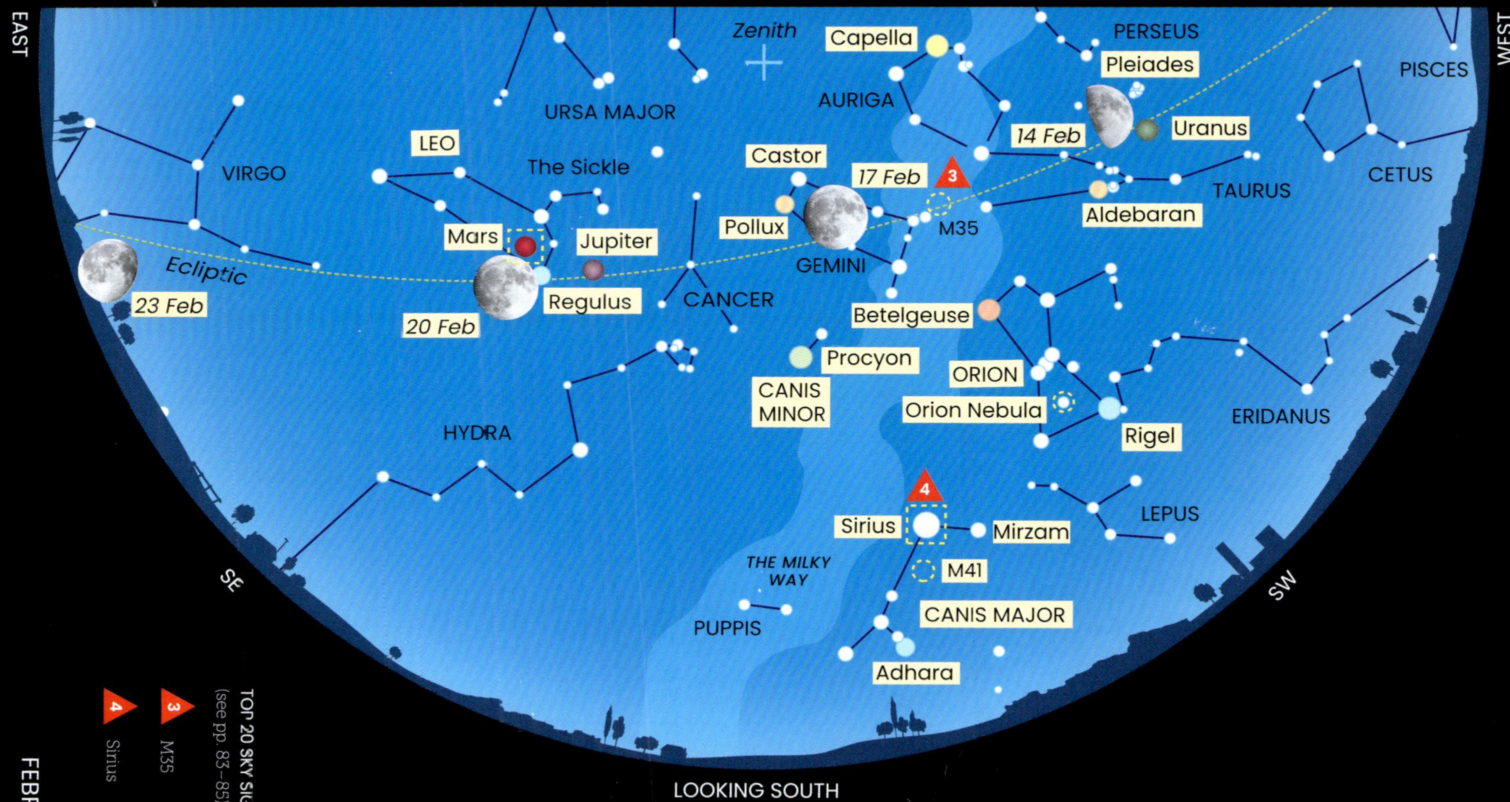

TOP 20 SKY SIGHTS
(see pp. 83–85)

3 M35

4 Sirius

Jupiter and **Mars** blaze together in the evening sky as the two planets both swing closest to the Earth this month. They're joined by the dazzling stars of winter: **Betelgeuse** and **Rigel** in **Orion**, escorted by glorious **Sirius** and – forming a giant arc above – **Procyon**, **Castor** and **Pollux**, **Capella** and the red giant **Aldebaran**.

FEBRUARY'S CONSTELLATION

Canis Major is the larger of **Orion**'s two hunting dogs – the other, naturally, being **Canis Minor** – and it's crowned by the Dog Star, brilliant **Sirius**. The brightest star in our skies, Sirius is one of the Sun's closest neighbours, only 8.6 light years away.

The beautiful star cluster **M41** is easily visible through binoculars, and even to the unaided eye. The Greek philosopher Aristotle, in 325 BC, referred to 'a cloudy spot' in Canis Major: if that was M41 it's the earliest surviving description of a deep-sky object.

FEBRUARY'S OBJECT

In 1854, the sight of the Red Planet in Leo inspired the poet Alfred, Lord Tennyson to describe how – in his poem 'Maud'– the eponymous heroine 'pointed to Mars/As he glow'd like a ruddy shield on the Lion's breast.' And **Mars** shines in the breast of Leo again this month, as it reaches its closest point to Earth.

Binoculars won't reveal any detail on this small world – just half the Earth's size – but a telescope will show you its dark surface markings, white frozen polar caps and the occasional white cloud or vast orange dust storm.

Space probes have discovered that diminutive Mars has truly astounding geology. Its enormous canyon, Valles Marineris, is 4000 kilometres long and 7000 metres deep – big enough to swallow the Alps. The giant dormant volcano, Olympus Mons, is three times the height of Mount Everest. And there's abundant evidence that the now-dry planet was once awash with rivers, lakes and maybe even oceans. In these watercourses, scientists are on the hunt for signs of microbiotic Martian life.

FEBRUARY'S TOPIC: A STAR'S LIFE

We think of the stars as being constant and unchanging, but that's just a result of our short human timescales. Look over a sweep of millions or billions of years, and the stars would change drastically before our eyes.

Let's start with the **Orion Nebula**, a stellar nursery packed with stars born only a couple of million years ago. The **Pleiades** is a family of stars just 100 million years old, now starting their adult life. Astronomers call this phase the

OBSERVING TIP

The brilliant winter constellations and planets are tempting us to spend plenty of time outside. So make sure you dress up warmly! Lots of layers are better than just a heavy coat, as they trap more air close to your skin. Heavy-soled boots with two pairs of socks stop the frost creeping up your legs. And a surprising amount of your body heat escapes through the top of your head, so complete your outfit with a beanie or – even better – a hat with ear flaps.

'main sequence', because you can rank these stars according to their mass: from brilliant hot heavyweights (like **Regulus**), through yellowish middling-mass stars (like the Sun), to dim lightweight red dwarfs so faint you'll need a telescope to see even the nearest.

As a massive star rips through its supplies of nuclear fuel, it expands immensely to become a red supergiant like **Betelgeuse**, and then explodes as a supernova. A Sun-like star lives for 10 billion years before it develops middle-age spread and turns into a red giant like **Aldebaran**, which is destined to puff off its outer layers as a 'planetary nebula', while its core survives as a dense white dwarf. The puny red dwarf stars live so long that none of them has expired in the lifetime of the Universe.

FEBRUARY'S PICTURE

This month's annular solar eclipse is visible from the uninhabited reaches of the South Pacific; but the path of annularity of the 2 October 2024 eclipse happened to cross right over the island of Rapa Nui (Easter Island). Josh Dury travelled there from the UK to image this spectacle of nature over the island's iconic moai statues.

For the Polynesian people who reached Rapa Nui around 800 years ago, a solar eclipse was a dark deed perpetrated by evil god Katiki, who lived in an extinct volcano and would emerge periodically to devour the Sun. In this case, his plan signally failed: not only was his mouth too small to encompass the entire solar orb, but – as this sequence shows – the Sun quickly got away again. But Katiki has plenty of time to refine his technique, as the next annular solar eclipse over Rapa Nui is not due for another 318 years.

Josh Dury composited this image from a series of exposures taken with a Sony A7S II camera plus Sigma 15 mm DG DN Diagonal Fisheye f/1.4 lens, and a Canon 5D Mark III camera equipped with a Sigma 150–600-mm Contemporary f/5-6.3 DG OS HSM lens.

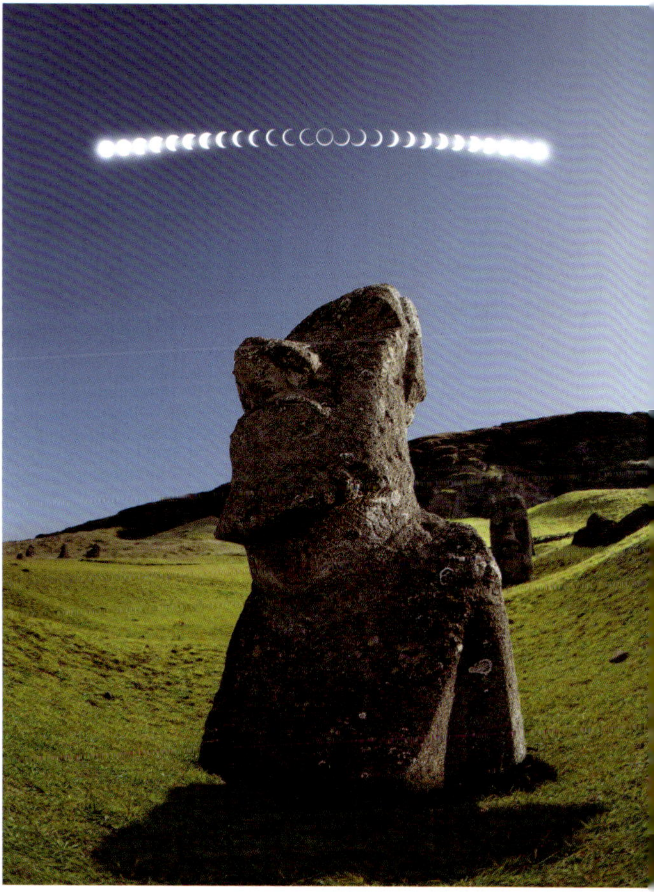

FEBRUARY'S CALENDAR

SUNDAY	MONDAY	TUESDAY	WEDNESDAY	THURSDAY	FRIDAY	SATURDAY
	1 Moon near Venus (am)	2 Moon near Venus (am)	3 Moon near Venus (am); Mercury E elongation	4	5	6 3.56 pm New Moon; annular solar eclipse
7	8 Moon near Mercury	9	10 Jupiter closest to Earth; Moon near Saturn	11 Jupiter opposition	12	13
14 7.58 am First Quarter Moon	15	16	17	18	19 Mars opposition; Moon near Jupiter	20 11.23 pm Full Moon; super-moon; Mars near Regulus and closest to Earth
21	22	23	24 Moon near Spica	25	26	27
28 5.16 am Last Quarter Moon near Antares (am)						

SPECIAL EVENTS

- **1 February, before dawn:** the crescent Moon lies to the right of Venus with Antares very nearby (Chart 2a).
- **2 February, before dawn:** a beautiful sight as Venus teams up with the Moon (Chart 2a).
- **3 February, before dawn:** the crescent Moon shines below Venus (Chart 2a).
- **3 February:** Mercury is at its greatest evening separation from the Sun (see Planet Watch).
- **6 February:** an annular eclipse of the Sun is visible from parts of the Pacific Ocean, South America, the Atlantic Ocean and western Africa.
- **8 February:** the crescent Moon hangs above Mercury.
- **10 February:** Jupiter is closest to the Earth at 652 million km (see Planet Watch).
- **11 February:** Jupiter lies opposite to the Sun.
- **19 February:** Mars lies opposite to the Sun (see Planet Watch).
- **19 February:** The Moon passes below Jupiter, with both Mars and Regulus lying nearby (Chart 2b).
- **20 February:** Mars is closest to the Earth at 101 million km.
- **20 February:** Mars passes near Regulus, with the Full Moon just below Regulus and Jupiter to the upper right (Chart 2b). Five celestial objects – the Sun, the Earth, the Moon, Mars and Regulus – are almost exactly in line.

FEBRUARY'S PLANET WATCH

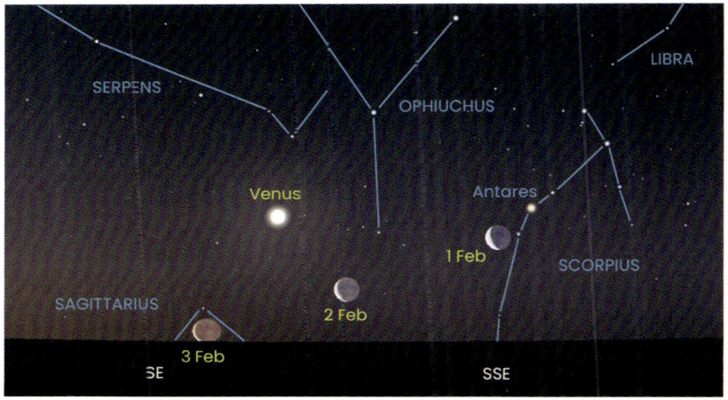

2a *1–3 February 6.30 am. Venus and the crescent Moon.*

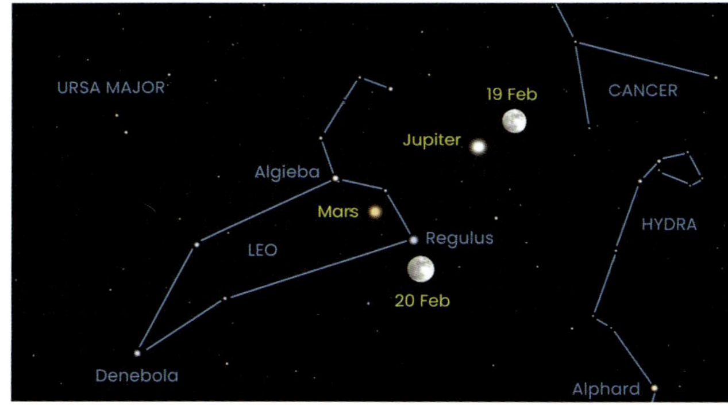

2b *19–20 February, 10 pm. The Moon with Jupiter, Mars and Regulus.*

- Look low on the western horizon early in the month for a glimpse of elusive **Mercury**, which reaches its greatest separation from the Sun on 3 February. Setting about 6.30 pm, the innermost planet fades from magnitude −0.6 to +1.0 during the first ten days of February. With binoculars, you can catch the thinnest crescent Moon above Mercury on 8 February.
- **Saturn** creeps from Cetus into Pisces this month. Shining at magnitude +0.8, the sixth planet is setting around 9 pm. The Moon is nearby on 10 February.
- Faint **Neptune**, in Pisces, glows at a mere magnitude +7.8 and sets about 8.30 pm.
- Setting around 2 am, **Uranus** (magnitude +5.7) lies in Taurus.
- **Jupiter** is the star-turn this month, at its closest point to Earth on 10 February and at opposition the next day. With a maximum brilliance of magnitude −2.6, Jupiter is outshone in the evening sky only by the Moon; located in Leo, the giant planet is visible all night long. Even with binoculars you can make out the four largest moons of its family of 101, and a small telescope reveals the bands of clouds wrapped around the gas giant. The Moon is nearby on 19 February (Chart 2b).
- Also in Leo, **Mars** (magnitude −1.3) is visible all night and heads rapidly upwards, past Regulus towards Jupiter. The Red Planet reaches opposition on 19 February and its closest point to Earth on 20 February, when the Full Moon lies nearby (Chart 2b). With a telescope, you can make out Mars's dark markings and white polar caps.
- Early birds can nab **Venus**, rising about 5 am. The Morning Star is resplendent at magnitude −4.2. The crescent Moon forms a lovely sight with Venus on the mornings of 1, 2 and 3 February (Chart 2a).

- The sky at 10 pm in mid-March, with Moon positions at three-day intervals either side of Full Moon.
- The star positions are also correct for 11 pm at the beginning of March, and 10 pm at the end of the month (after BST and IST begin).
- The planets move slightly relative to the stars during the month.

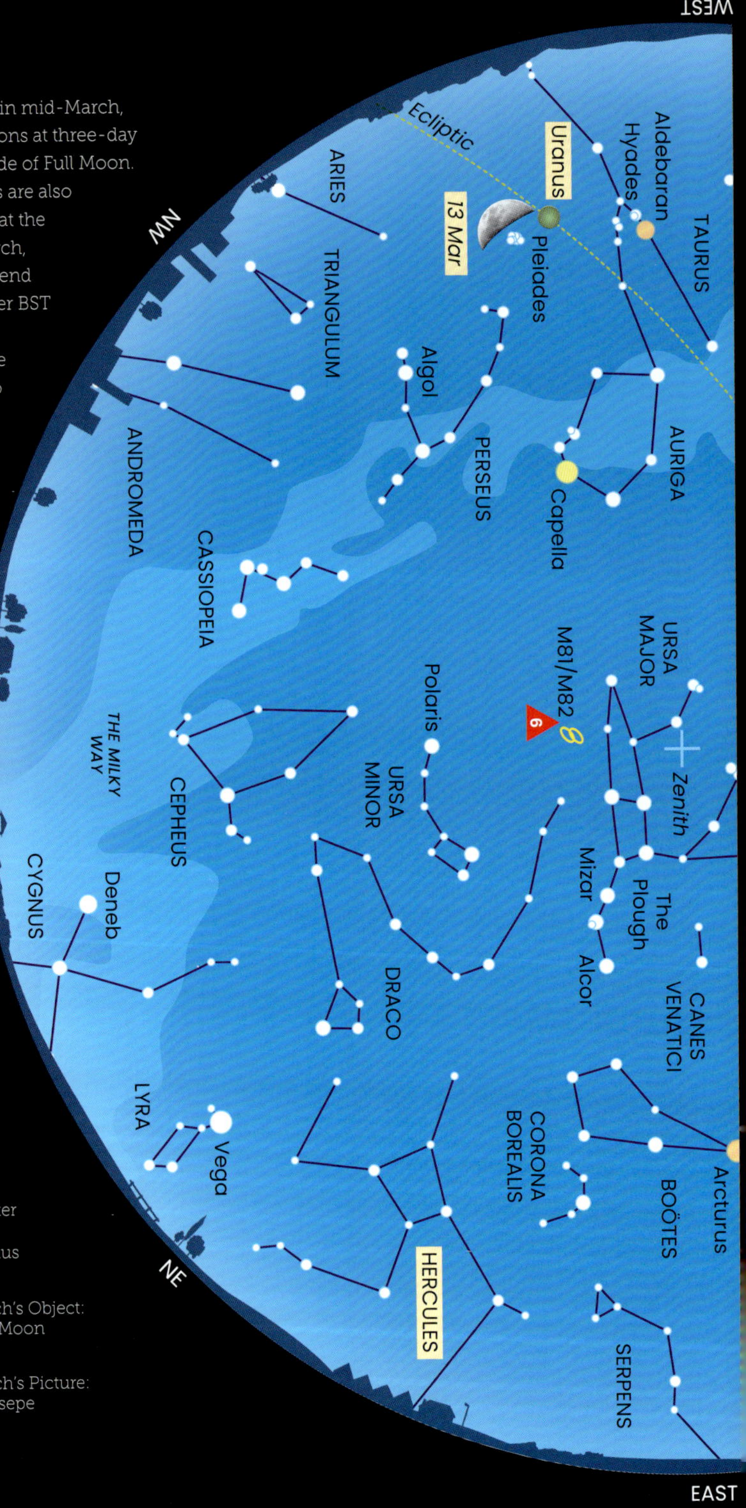

- Mars
- Jupiter
- Uranus
- March's Object: The Moon
- March's Picture: Praesepe

MARCH

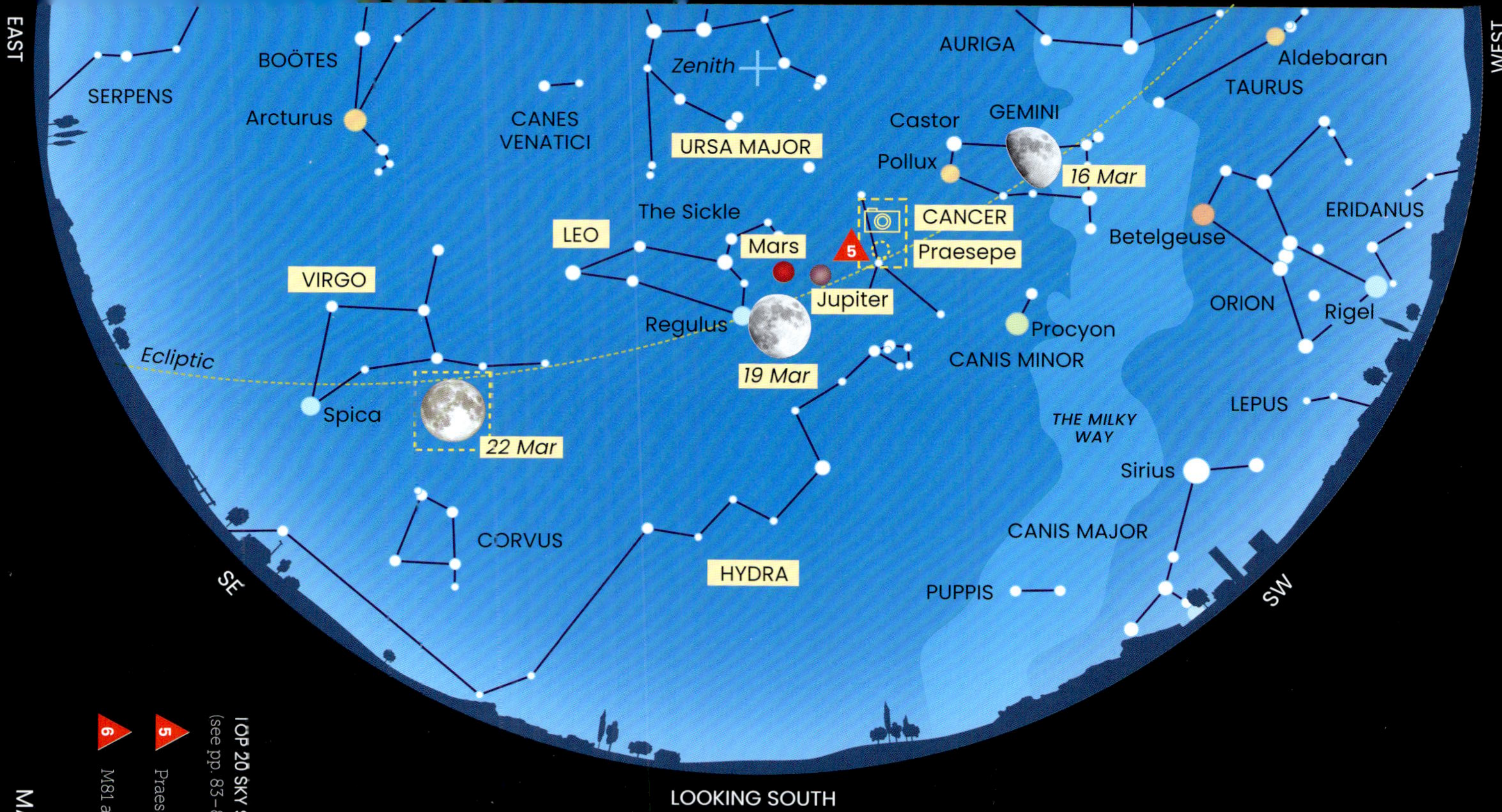

TOP 20 SKY SIGHTS
(see pp. 83–85)

- 5 Praesepe
- 6 M81 and M82

The nights are bookended by brilliant planets: **Jupiter** and **Mars** in the evening, and Venus before dawn. The large spring constellations **Leo**, **Virgo** and **Hydra** are well on display in the southern sky.

MARCH'S CONSTELLATION

Although it's well known as one of the constellations of the Zodiac, **Cancer** is easy to overlook because its stars are so faint that city lights completely drown them out.

In legend, Cancer attempted to nip **Hercules**'s foot during his altercation with the multi-headed monster **Hydra**. When Hercules crushed the crustacean under his foot, the goddess Juno took pity on the poor beast and placed it in the sky.

The constellation's main claim to fame is the star cluster **Praesepe**. Known since ancient times – the Greek astronomer Ptolemy called it 'a nebulous mass in the breast of the Crab' – this swarm of stars is a magnificent sight through binoculars or a small telescope (see Picture).

MARCH'S OBJECT

Our nearest celestial companion, the **Moon**, lies a mere 384,400 kilometres away – 400 times closer than the Sun. At 3,476 kilometres across, the Moon is so large compared to our world that astronomers often call the Earth–Moon system a double planet.

The Moon is the only night-time object on which we can see any details with the unaided eye: a pattern of dark splodges making up the face of the 'Man in the Moon', or – according to other cultures – the outline of a rabbit, a wolf or an old man carrying a bundle of sticks.

As the Moon moves round the Earth once a month (a word that was originally a 'moonth'), the amount of its surface lit by the Sun changes, from a narrow crescent to a completely illuminated Full Moon, and back to a crescent. A couple of times per year (on average), the Moon moves into the Earth's shadow and we see it fade from sight in a lunar eclipse.

Though the Moon looks brilliant – at magnitude −12.5 the Full Moon is thousands of times brighter than all the stars and planets combined – the lunar surface is actually as dark as an asphalt road. If it were covered with ice, like Saturn's moon Enceladus, the Full Moon would truly dazzle us, and its light would make the night sky shine a deep shade of blue.

MARCH'S TOPIC: PTOLEMY (c.100–c.170)

Arguably the most influential astronomer of all time was Claudius Ptolemaeus (known colloquially as Ptolemy), a Roman citizen living near the great library of Alexandria in ancient Egypt. Today, we

OBSERVING TIP

This is the ideal time of year to tie down the main compass directions, as seen from your observing site. North is easy – just latch onto Polaris, the Pole Star, using the familiar stars of the Plough (see May's Constellation). And at noon, the Sun is always in the south. But the useful extra in March is that we hit the Spring Equinox, when the Sun rises due east, and sets due west. Memorise those positions relative to a tree or house around your horizon, and you'll know which way to orientate the Star Charts.

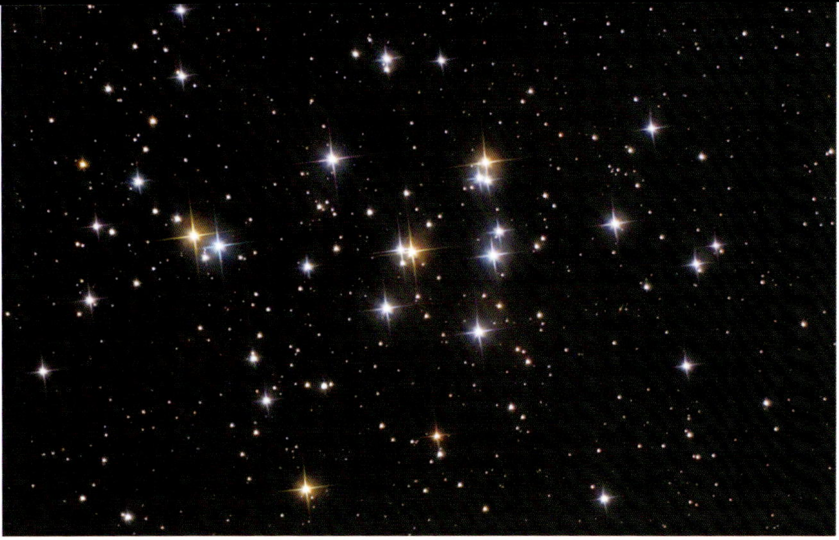

still use the 48 constellations that he listed – from glorious Orion to obscure Triangulum – and many of our star names are Arabic translations of Ptolemy's original monikers, such as Fomalhaut ('the mouth of the southern fish').

But Ptolemy is mainly remembered for his theory of the planetary motions, which lasted 1400 years. In the Ptolemaic system, the planets move around the Earth, each on a small circle (an epicycle) whose centre travels on a circular orbit around us. Not until 1543, when Nicolaus Copernicus shifted the Sun to the centre of the Solar System, was this theory overthrown.

In fact, previous Greek astronomers had devised most of his constellations and even the theory of epicycles, but their original writings were lost. Ptolemy struck lucky: his great astronomical tome survived, and was passed down to Arabic astronomers and then on to Europe. Its Arabic name says it all: the Almagest, meaning 'the greatest'.

MARCH'S PICTURE

A celestial jewel box of diamonds and rubies scattered on black velvet, the star cluster Praesepe is a magnificent sight through a small telescope. The diamonds in Bill McSorley's evocative image are stars like Sirius – hotter and brighter than the Sun – while the rubies are more massive stars that have swollen up to become red giants.

Surrounded by the street-lights of Leeds, Bill McSorley employed a Sky-Watcher LP light pollution filter on his 150-mm Sky-Watcher Explorer 150P Newtonian reflector to capture the bright lights of Praesepe against a black background. He took 9 × 300-seconds (45 minutes total) exposures on a QHY 8L cooled one-shot colour camera, and processed the image in PixInsight and Photoshop.

To the naked eye, Praesepe appears as a faint misty patch in Cancer (see Constellation). The traditional name means 'the manger', but even a pair of binoculars will resolve Praesepe into a swarm of stars – spread over an area three times the size of the Moon – which has inspired its popular nickname the Beehive Cluster.

The latest surveys reveal over 1000 stars in Praesepe, stretching out over 70 light years. Lying some 600 light years away, Praesepe was born from a dense gas cloud about 600 million years ago, making it a youngster on the cosmic stage.

MARCH'S CALENDAR

SUNDAY	MONDAY	TUESDAY	WEDNESDAY	THURSDAY	FRIDAY	SATURDAY
	1	2	3	4 Moon near Venus (am)	5 Moon near Venus (am)	6
7	8 9.29 am New Moon	9	10 Moon near Saturn	11	12	13 Moon occults Pleiades
14 Moon occults Pleiades (am)	15 5.25 pm First Quarter Moon	16	17 Mercury W elongation	18 Moon near Jupiter	19 Moon near Mars	20 Spring Equinox
21	22 10.44 am Full Moon	23 Moon near Spica	24	25	26	27 Moon near Antares (am)
28 Moon near Antares (am); BST and IST begin (am)	29	30 1.54 am Last Quarter; Mars nearest to Jupiter	31			

SPECIAL EVENTS

- **4 March, before dawn:** low in the brightening eastern sky, Venus lies to the left of the crescent Moon.
- **5 March, before dawn:** a thin crescent Moon is located to the left of brilliant Venus in the dawn glow.
- **10 March:** a thin crescent Moon hangs next to Saturn.
- **Night of 13/14 March, 9.30 pm–0.30 am:** the Moon passes in front of the Pleiades and occults half its brightest members (Chart 3a). With binoculars or a telescope, watch the stars suddenly blink out behind the Moon's dark edge.
- **17 March:** Mercury reaches its greatest morning separation from the Sun (see Planet Watch).
- **18 March:** to the left of the Moon lie Jupiter, Mars and Regulus (Chart 3b).
- **19 March:** you'll find a striking sight in the evening sky tonight, as the Moon forms a tight group with Jupiter, Mars and Regulus (Chart 3b).
- **20 March, 9.25 pm:** the Spring Equinox, when day and night are equal.
- **28 March, 1.00 am:** British Summer Time and Irish Standard Time start – put your clocks forward tonight.
- **30 March:** Mars reaches its closest approach to Jupiter (4.3 degrees) before backing away again.

MARCH'S PLANET WATCH

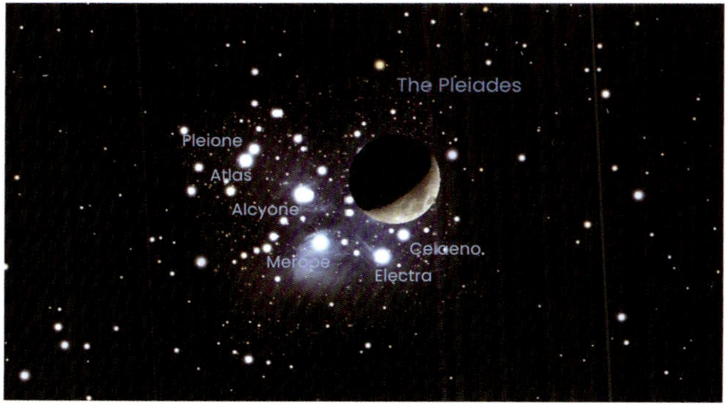

3a *13 March, 11 pm. The Moon occults the Pleiades.*

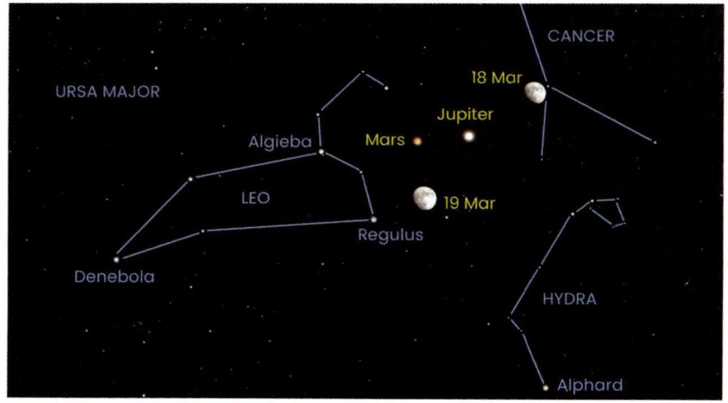

3b *18–19 March, 9 pm. The Moon with Jupiter, Mars and Regulus.*

- At the beginning of March, you'll find **Saturn** low in the west after sunset, shining at magnitude +0.8 in Pisces and setting at 8.30 pm. The crescent Moon passes to the right of the ringworld on 10 March. But Saturn is sinking into the twilight glow, and has disappeared from sight by the end of the month.
- **Uranus** (magnitude +5.8) is in Taurus, and sets about 0.30 am.
- Visible throughout the night, brilliant **Jupiter** is dominating the nocturnal view as it shines at magnitude −2.5 on the border of Leo and Cancer. The Moon lies to the right of Jupiter on 18 March (Chart 3b).
- Following closely on Jupiter's heels – and also visible all night – is **Mars**. At the start of March, the Red Planet is steaming through Leo towards Jupiter, but its momentum slows and Mars grinds to a halt 4.3 degrees from Jupiter at the end of the month. During March, Mars fades precipitously as the Earth pulls away, from magnitude −1.1 to −0.4. The Moon is nearest to Mars on 19 March, when there's a gorgeous tableau that also includes Jupiter and Regulus (Chart 3b).
- Rising at 5 am, **Venus** is easy to spot as the Morning Star at the beginning of March: the crescent Moon is nearby on the mornings of 4 and 5 March. But it's dropping down into the dawn twilight and becomes hard to see by the end of the month, despite shining at magnitude −4.0.
- **Neptune** is lost in the Sun's glare this month, as is **Mercury** even though it reaches its greatest morning separation from the Sun on 17 March.

- The sky at 11 pm in mid-April, with Moon positions at three-day intervals either side of Full Moon.
- The star positions are also correct for midnight at the beginning of April, and 10 pm at the end of the month.
- The planets move slightly relative to the stars during the month.

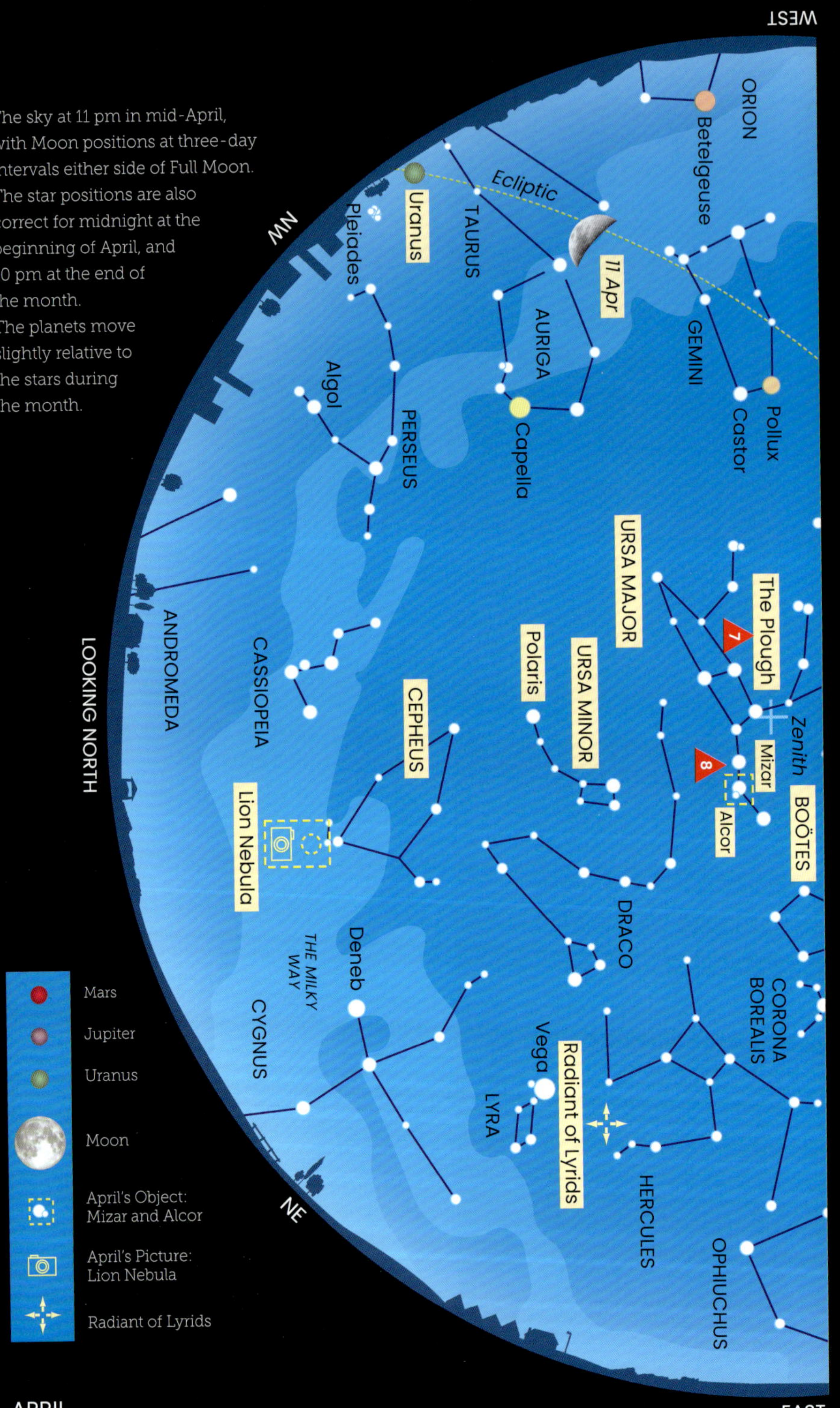

APRIL

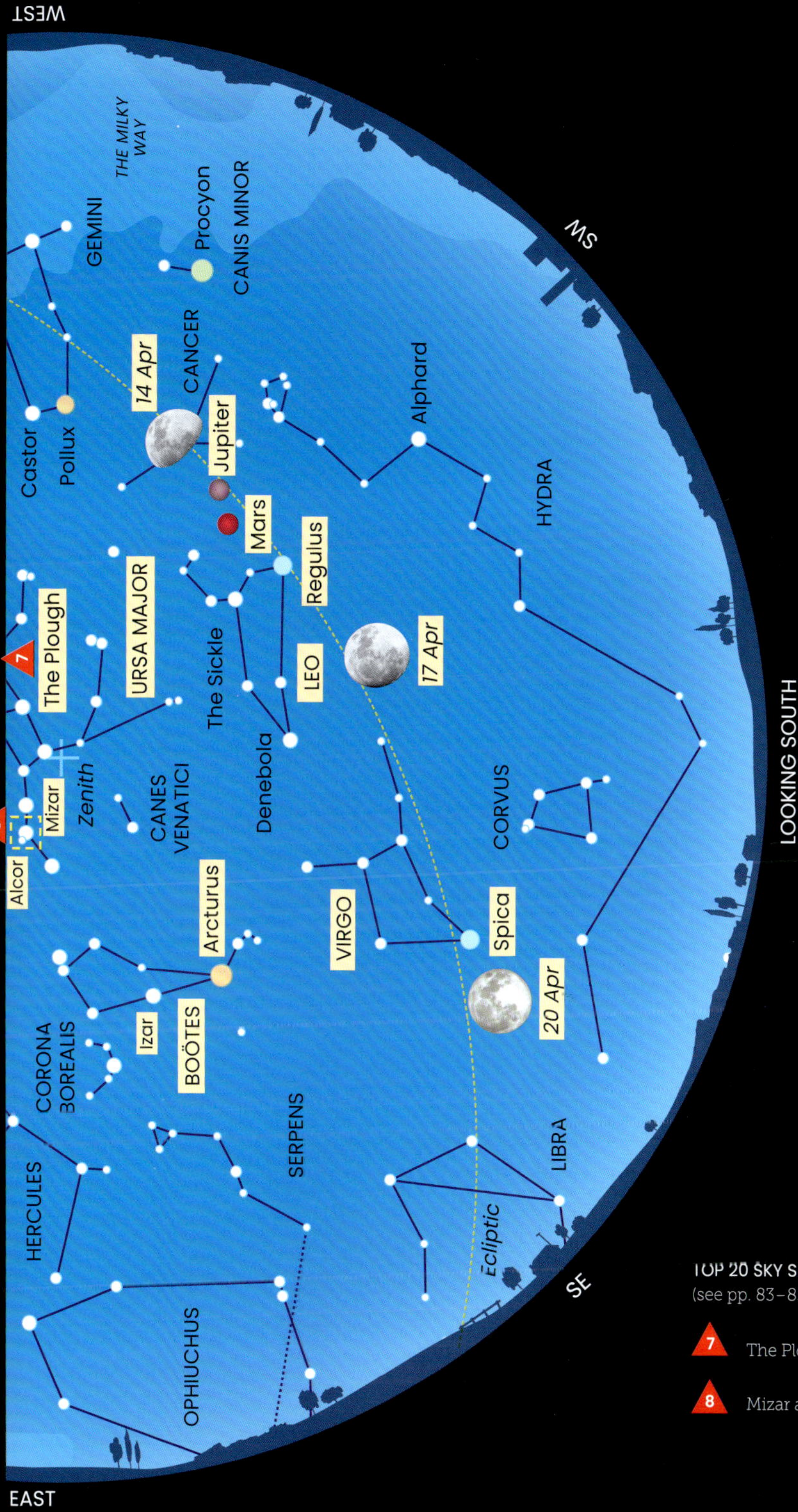

TOP 20 SKY SIGHTS
(see pp. 83–85)

7 The Plough

8 Mizar and Alcor

Jupiter and **Mars** are beacons in the western sky, balanced – to the south-east – by three bright stars that are the principal players in the spring constellations. Leading the way is **Regulus** in **Leo**, with **Virgo**'s chief star **Spica** to the lower left, and orange **Arcturus** in **Boötes** lying above.

APRIL'S CONSTELLATION

The kite-shaped constellation **Boötes**, the Herdsman, was mentioned in Homer's *Odyssey*, and its name refers to the fact that Boötes seems to herd the northern stars as they circle the Pole Star, **Polaris**. In particular, the brightest star **Arcturus** ('bear-driver') apparently drives the two bears (**Ursa Major** and **Ursa Minor**) around the sky.

Arcturus is the fourth brightest star in the whole sky, and it's the most brilliant star you can see on April evenings. A red giant star in its old age, Arcturus lies 37 light years from us, and shines 170 times more brilliantly than the Sun.

Another fine sight in Boötes is **Izar**, meaning 'the belt'. Through a good telescope, it appears as a gorgeous double star – one star yellow and the other blue.

OBSERVING TIP

It's always fun to search out 'faint fuzzies' in the sky – star clusters, nebulae and galaxies. But don't even think of observing these objects around the time of Full Moon, as its light will drown them out. You'll have the best views near New Moon: a period astronomers call 'dark of Moon'. When the Moon is bright, focus on planets, bright double stars and, of course, the Moon itself.

APRIL'S OBJECT

To spot the most famous pair of stars in the sky – **Mizar** (magnitude +2.0) and **Alcor** (magnitude +4.0) – home in on the 'kink' in the tail of Ursa Major (the Great Bear), which if you prefer is the handle of **the Plough**.

This is one of the few double stars you can resolve with the naked eye: most other star pairs are so tight that you need a telescope. Although some astronomers have mooted that Mizar and Alcor just happen to lie along the same line of sight, the most recent measurements put them at such similar distances – 82 light years for Alcor and 83 for Mizar – that they probably are in a gravitational embrace.

Mizar has a companion that's visible through a telescope, and both these stars are actually very close doubles, making four stars in all. And Alcor has a faint companion – so there's a total of six stars making up this amazing system.

APRIL'S TOPIC: BLACK HOLES

A black hole is science fiction become science fact: an intense gravitational well in space, with a force so intense that anything that falls in – even a ray of light – can never escape.

The first evidence came when astronomers found that some ordinary stars are whirling around an object that we can't detect directly. When they weighed the dark companion, they found it was too massive to be a collapsed star (like a white dwarf or neutron star). Instead, it had to be the core of a star that had exploded

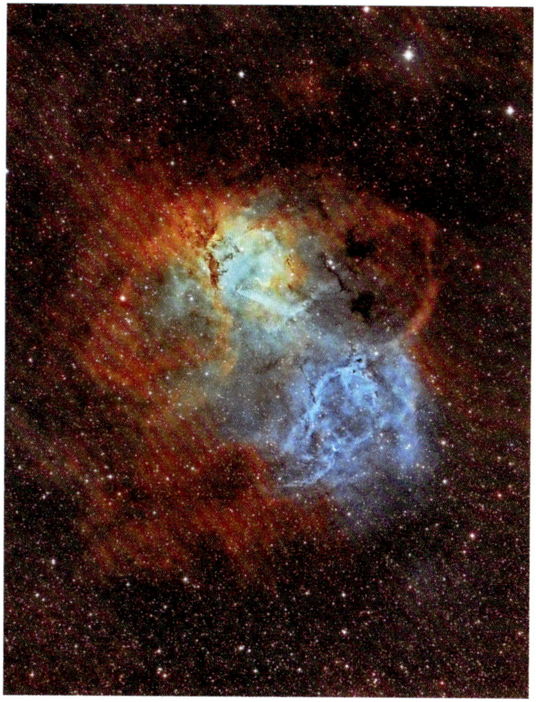

Like other modern big-game hunters, Sara Wager shot the Lion Nebula with only a camera – in this case, a QSI 683 cooled CCD camera attached to a Takahashi FSQ85 85-mm f/5.3 refractor. Through Astrodon 3nm filters, she collected 35 × 1800-second exposures in hydrogen alpha, oxygen ([OIII]) and sulphur (SII) light, for a total exposure time of 52 hours 30 minutes.

as a supernova, collapsed inwards to become a black hole around ten times heavier than the Sun.

Researchers have also picked up gravitational waves – ripples in space itself – coming from further out in space. These vibrations end in a 'chirp', marking the moment that two orbiting black holes coalesce into a single black hole. These mergers typically involve black holes about 40 times the Sun's mass.

And there's overwhelming evidence for supermassive black holes in the centres of galaxies. Stars at the core of our own Milky Way are speeding in tight orbits around a black hole as heavy as 4 million Suns. In the past few years, astronomers using radio telescopes spread across the Earth have obtained images of the central black hole in the Milky Way – and also in the giant galaxy M87 – silhouetted against a ring of glowing gas that's swirling around the black hole itself.

APRIL'S PICTURE

Let your imagination loose as you view this cosmic gas cloud in **Cepheus**, and you can see why it's called the **Lion Nebula**: orange tail on the right, four bluish legs in the middle, and a wonderful maned head on the left.

Officially known as Sh2-132 (from a catalogue of faint nebulae compiled by American astronomer Stewart Sharpless), the lion covers an area larger than the Full Moon – but it's so faint you'll be hard pushed to see it though a telescope. Its true majesty only appears in long-exposure photographs, like this image from Sara Wager, a British astrophotographer viewing the heavens under the clear dark skies in Spain.

The Lion Nebula strides through the Perseus Arm of our galaxy, around 10,400 light years away. Its gas clouds surround the hot young stars of the Cepheus OB1 Association, whose ultraviolet radiation excites hydrogen gas to glow red and oxygen atoms to fluoresce in green.

APRIL'S CALENDAR

SUNDAY	MONDAY	TUESDAY	WEDNESDAY	THURSDAY	FRIDAY	SATURDAY
				1	2	3
4	5	6	7 0.51 am New Moon	8	9	10
11	12	13 11.56 pm First Quarter Moon	14 Moon near Praesepe	15 Moon near Praesepe (am); Moon near Jupiter and Mars	16 Moon near Regulus	17
18 Moon near Spica	19 Moon near Spica	20 11.27 pm Full Moon	21	22 Lyrids	23 Lyrids (am)	24 Moon near Antares (am)
25	26	27	28 9.18 pm Last Quarter Moon	29	30	

SPECIAL EVENTS

- **Night of 14/15 April:** the Moon grazes the lower edge of Praesepe (Chart 4a).
- **15 April:** the Moon lies between Jupiter (right) and Mars (left), with Regulus further to the left again (Chart 4b).
- **Night of 22/23 April:** shooting stars from the **Lyrid meteor shower** – debris from Comet Thatcher burning up in the Earth's atmosphere – reach their peak tonight, but the show is spoilt this year by bright moonlight.
- **24 April, early hours:** the Moon skims just below Antares; observers in southern Africa will see the Moon occult the red giant.

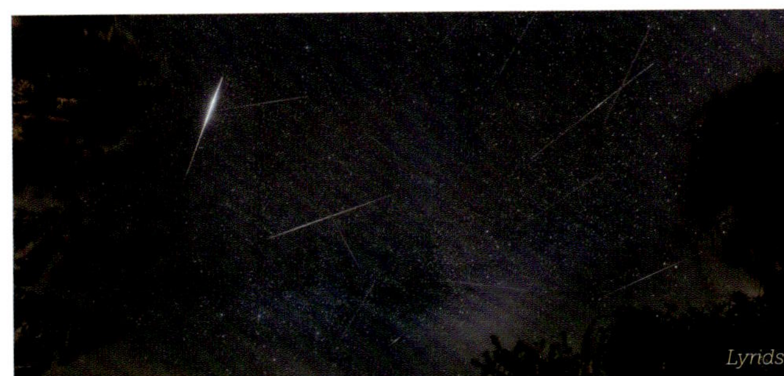

Lyrids

APRIL'S PLANET WATCH

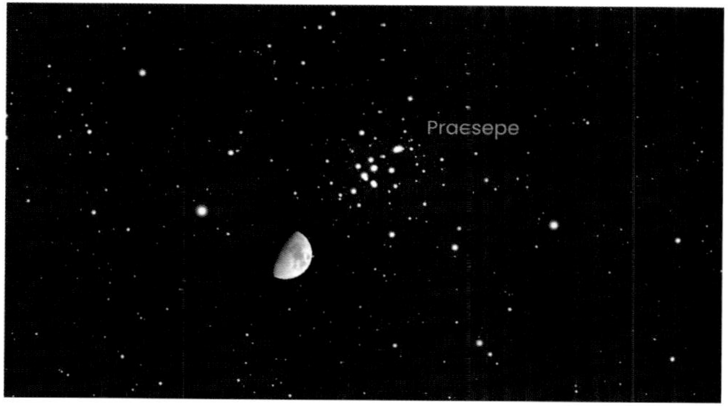

4a *15 April, 1 am. The Moon passes under Praesepe.*

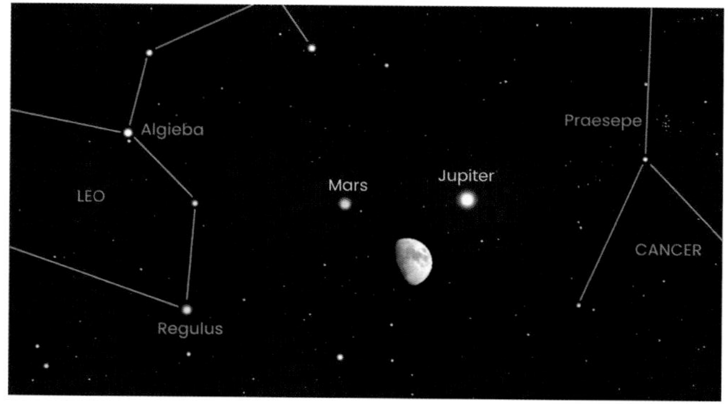

4b *15 April, 9 pm. The Moon with Jupiter, Mars and Regulus.*

- Second only to the Moon in brilliance at magnitude −2.2, **Jupiter** lies in Cancer this month and sets around 4 am. With few other planets on view, take the chance to study Jupiter and its moons.
- From night to night, binoculars will show you Jupiter's four brightest moons in a constantly changing dance as they orbit the planet: little Io speeds around Jupiter in less than two days, while the outermost, Callisto, has an orbital period of 17 days – all far shorter than the month that our Moon takes to orbit the Earth.
- A telescope reveals bands in Jupiter's clouds, and the Great Red Spot – a storm larger than the Earth that's been raging for over a century.
- **Mars** is almost ten times fainter than Jupiter at magnitude 0.0: lying in the neighbouring constellation of Leo, it's now travelling away from Jupiter and back towards Leo's leading light, Regulus. Mars, too, is setting about 4 am, but you'll need a reasonable-sized telescope to see any detail on its dusty surface.
- The Moon forms a pretty tableau with these planets on 15 April (Chart 4b).
- **Uranus** (magnitude +5.8) lies in Taurus and sets around 11.30 pm. This is your last chance to spot the seventh planet in the evening sky in 2027; next month it disappears into the twilight glow. You'll find Uranus – most easily in binoculars – halfway between the two bright star clusters of Taurus, the Pleiades and the Hyades – look for a 'star' that's moving upwards night after night.
- **Mercury**, **Venus**, **Saturn** and **Neptune** are too close to the Sun to be seen in April.

The sky at 11 pm in mid-May, with Moon positions at three-day intervals either side of Full Moon. The star positions are also correct for midnight at the beginning of May, and 10 pm at the end of the month. The planets move slightly relative to the stars during the month.

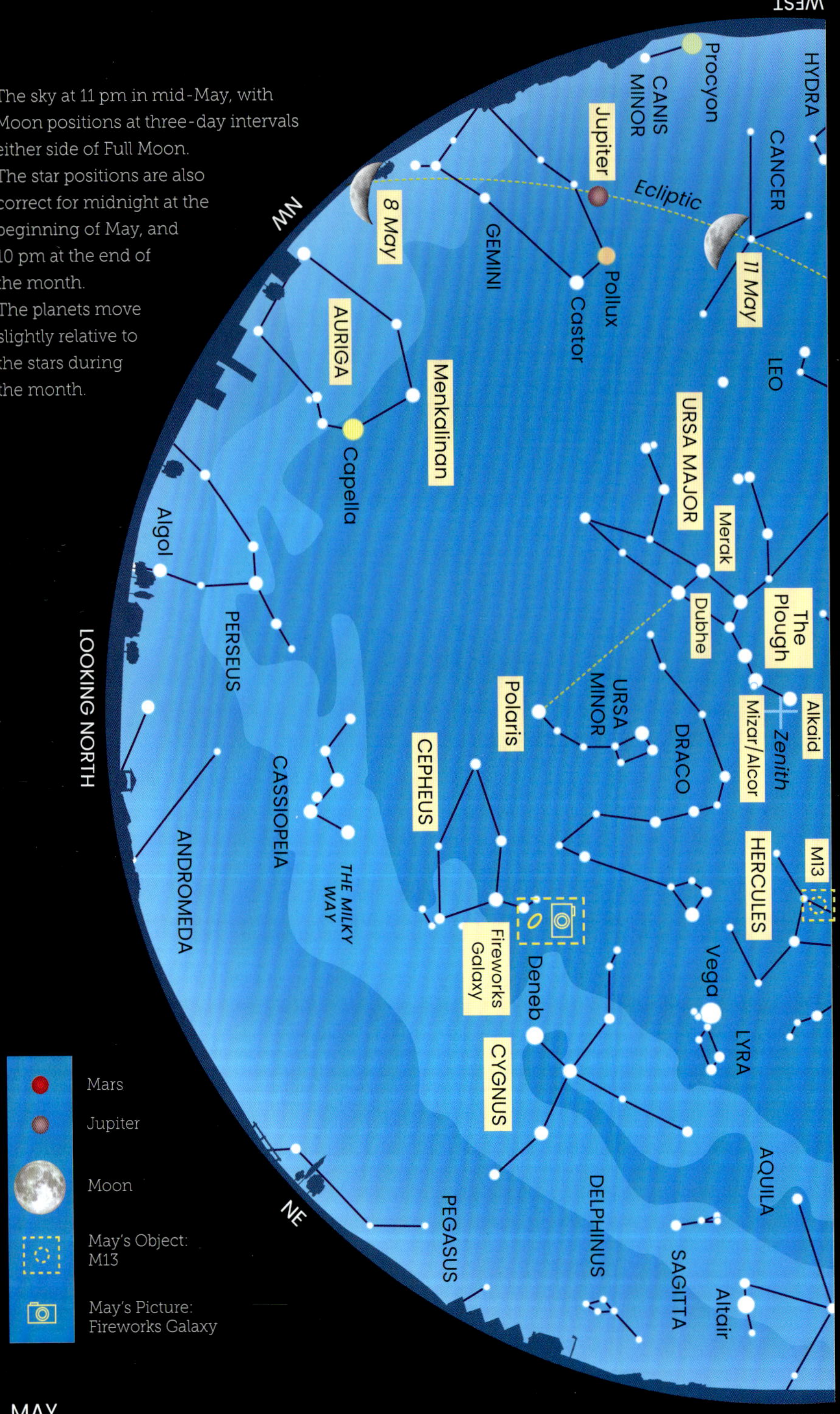

MAY

MAY

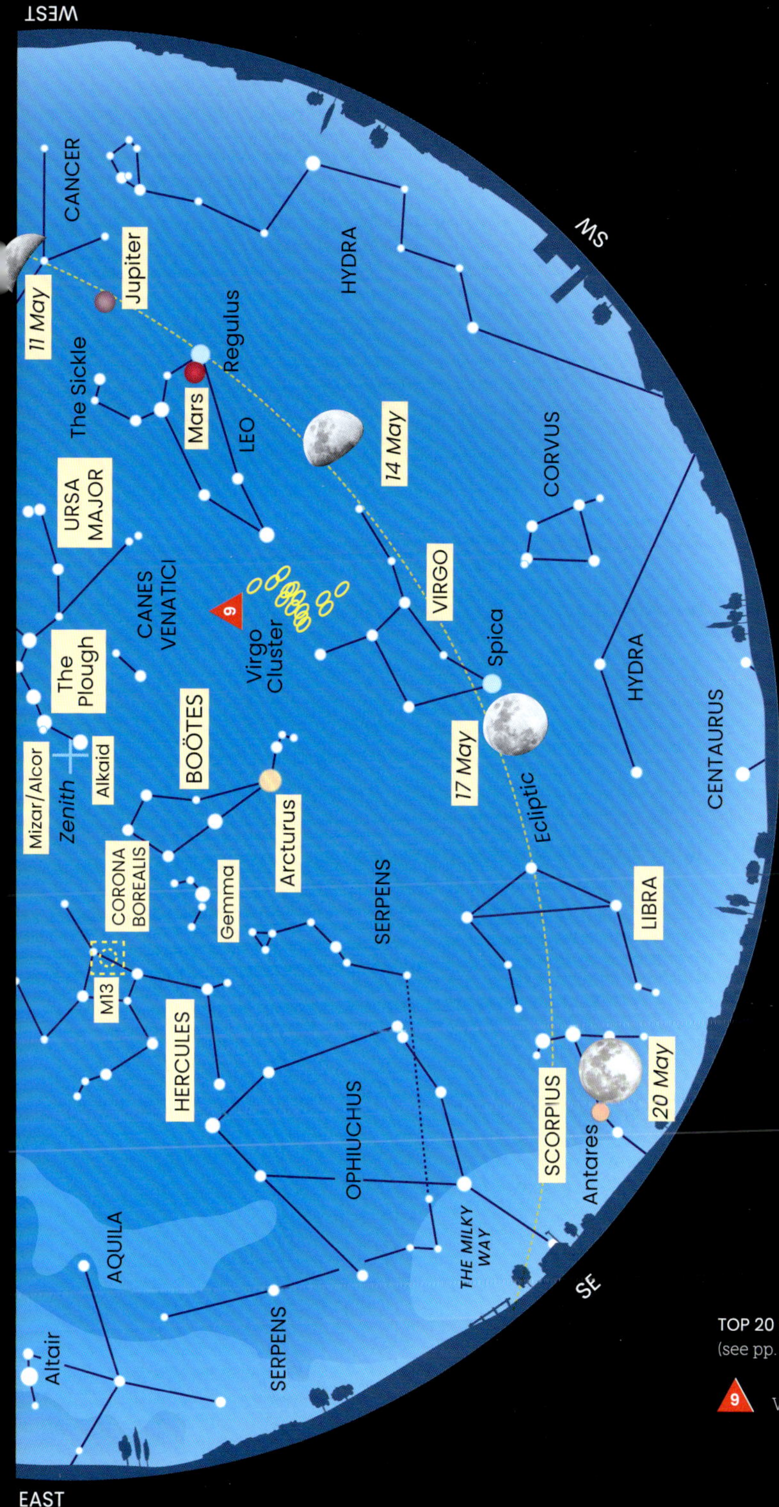

TOP 20 SKY SIGHTS
(see pp. 83–85)

9 Virgo Cluster

The long-standing duo of **Mars** and **Jupiter** is joined this month by their shy sibling Mercury, putting on its best evening appearance of the year. We can also enjoy the early summer constellations: **Boötes**, with the little circlet of **Corona Borealis** to its left, along with **Scorpius**, **Libra** and **Virgo** to the south.

MAY'S CONSTELLATION

Ursa Major, the Great Bear, is a constellation with deep roots. This star pattern may date back 30,000 years, when Palaeolithic people in Siberia erected shrines to cave bears. The bear tradition was then exported both to southern Europe and to the New World.

The seven brightest stars are often called **the Plough** or, in North America, the Big Dipper. You can use its two end stars – **Merak** and **Dubhe** – to locate **Polaris**, the Pole Star. In the middle of the bear's tail you'll find the celebrated double star **Mizar** and **Alcor** (see April's Object).

Unlike most constellations, the majority of the stars in the Plough lie at the same distance and were born together as members of the Ursa Major Moving Group, which also includes **Menkalinan** in **Auriga** and **Gemma** in **Corona Borealis**. The exceptions are Dubhe and **Alkaid**, and – as these two stars go off on their own paths – the shape of the Plough will gradually change.

OBSERVING TIP

It's best to view your favourite objects when they're well clear of the horizon. If you observe them low down, you're looking through a large thickness of the atmosphere – which is always shifting and turbulent. It's like trying to observe the outside world from the bottom of a swimming pool! This turbulence makes the stars appear to twinkle. Low-down planets also twinkle, although to a lesser extent, because they subtend tiny discs, and aren't so affected.

MAY'S OBJECT

At the darkest part of the night, you may spot a faint fuzzy patch way up in the constellation **Hercules**. Through binoculars, it appears as a gently glowing ball of light. With a telescope, you can glimpse its true nature: a cluster of half a million stars, swarming together in space.

This wonderful object is known as **M13**, because it was the 13th entry in the catalogue of fuzzy objects recorded by the 18th-century French astronomer Charles Messier. We now classify M13 as a 'globular cluster'. These great round balls of stars are among the oldest objects in our Milky Way Galaxy, dating back to its birth some 13 billion years ago.

In 1974, radio astronomers sent a message towards M13, hoping to inform the inhabitants of any planet there of our existence. There's only one problem: M13 lies so far away that we wouldn't receive a reply until AD 48,000!

MAY'S TOPIC: POLYNESIAN ASTRONOMY

The sable skies over the Pacific Ocean are crammed with stars, right down to the horizon. To the ancient Polynesians navigating the vast empty reaches, these

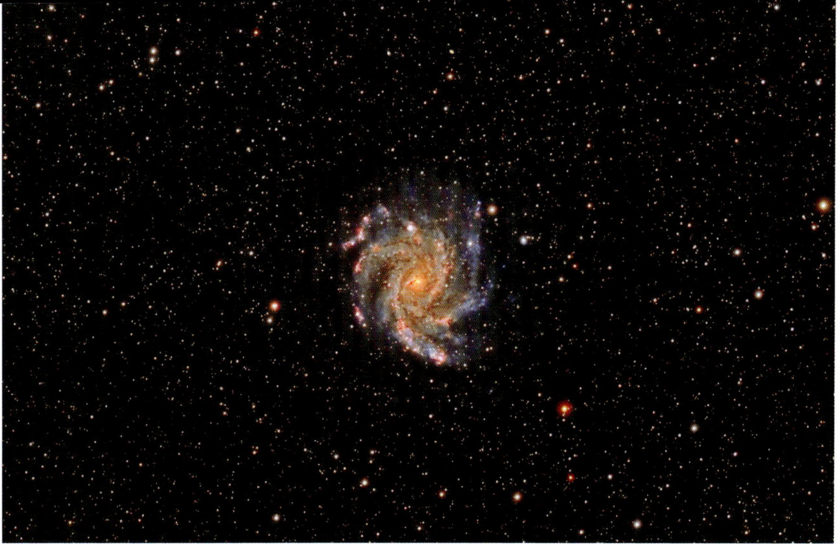

From the north-east Yorkshire coast, Steve Doody captured the Fireworks Galaxy with a 132-mm William Optics 132 FLT f/6.9 refractor, equipped with an Atik 460EX monochrome CCD camera cooled to −15° C. Using 180-second exposures, he took 18 frames through a red filter, 17 in green, 10 in blue and 20 luminance frames; plus 32 × 300-second frames through a hydrogen alpha filter, for a total acquisition time of 3 hours 40 seconds. He processed the image in PixInsight.

stars formed a huge celestial compass. A Polynesian navigator memorised where 220 of the brightest stars (along with the Pleiades) rose and set around the horizon. Once they were at sea, they could use this 'star compass' to work out in which direction they were sailing.

The Southern Cross (the constellation Crux) was also an essential navigational aid: when it stood upright in the sky, a line joining the top and bottom stars pointed due south.

The final tool in their navigational kit was to check the altitude of bright stars. The position of the stars changes as you sail north or south, so that Sirius, for example, passes straight overhead when you are in Tahiti, and Arcturus is the zenith star for Hawai'i. In 1976, a reconstructed Polynesian voyaging canoe – *Hōkūle'a* (meaning 'Arcturus') – was sailed 4000 kilometres from Hawai'i to Tahiti using only the traditional navigation techniques.

MAY'S PICTURE

Straddling the border of **Cepheus** and **Cygnus**, NGC 6946 is a small galaxy with a big reputation: over the past 110 years, ten supernovae have exploded in its spiral arms – a rate of stellar suicide that's ten times higher than we experience in the Milky Way. While most of these supernovae were tracked down through professional surveys, the star that exploded in 2008 was discovered by British amateur astronomer Ron Arbour.

Not surprisingly, NGC 6946 has attracted the moniker the **Fireworks Galaxy**. And it is a striking sight in long-exposure images, like this portrait taken by Steve Doody. But for the visual observer, the galaxy is surprisingly shy: its light is spread out over a comparatively large area, and you'll need really dark skies and a moderate telescope to spot it – and maybe bag the next supernova yourself!

MAY'S CALENDAR

SUNDAY	MONDAY	TUESDAY	WEDNESDAY	THURSDAY	FRIDAY	SATURDAY
30	31					1
2	3	4	5	6 11.58 am New Moon	7 Eta Aquarids (am)	8
9	10	11	12 Moon near Jupiter	13 5.44 am First Quarter Moon near Mars and Regulus	14 Mars near Regulus	15
16	17 Moon near Spica	18	19	20 11.59 am Full Moon near Antares	21	22
23	24	25	26	27	28 2.58 pm Last Quarter Moon; Mercury E elongation	29

SPECIAL EVENTS

- **7 May, early hours:** the maximum of the Eta Aquarid meteor shower should provide an excellent chance for catching shooting stars, as the Moon isn't in the way. These speedy meteors are tiny pieces of Halley's Comet burning up in Earth's atmosphere.

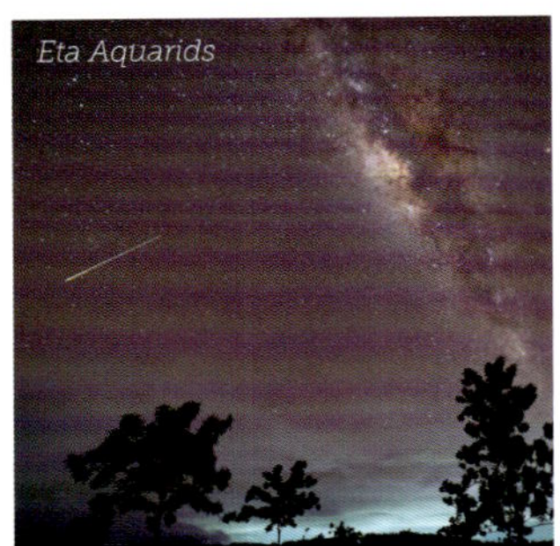
Eta Aquarids

- **12 May:** the bright 'star' that can be observed above the Moon is Jupiter, with Mars and Regulus to the left (Chart 5a).
- **13 May:** immediately above the Moon, you'll find Regulus and Mars, with Jupiter to the right (Chart 5a).
- **14 May:** Mars passes just above Regulus.
- **28 May:** Mercury reaches its greatest evening separation from the Sun (see Planet Watch).

MAY'S PLANET WATCH

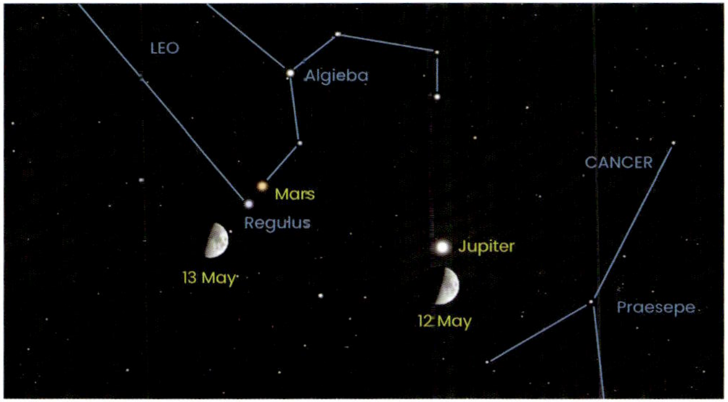

5a *12–13 May, 10.30 pm. The Moon with Jupiter, Mars and Regulus.*

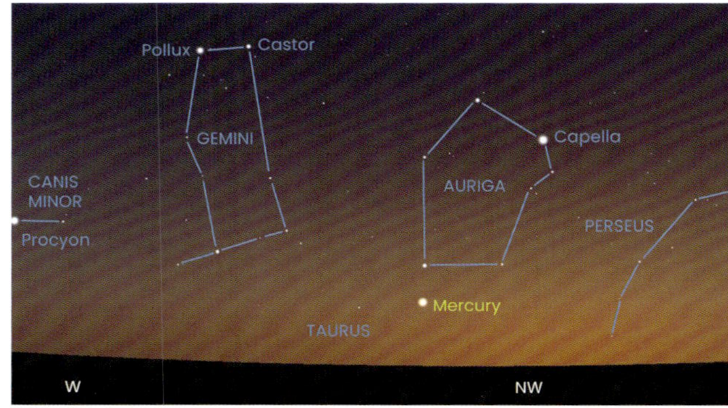

5b *20 May, 10 pm. Mercury is prominent in the dusk sky.*

- **Jupiter** blazes in Leo at magnitude –2.0, and sets around 2.30 am. The Moon passes just below the giant planet on 12 May (Chart 5a).
- To the left of Jupiter you'll find **Mars**, ten times fainter at magnitude +0.5. During May, the Red Planet is heading leftwards in the sky, away from Jupiter, and on 14 May it flies closely over Leo's leading light, Regulus. The previous night, 13 May, the First Quarter Moon lies just below Regulus and Mars (Chart 5a).

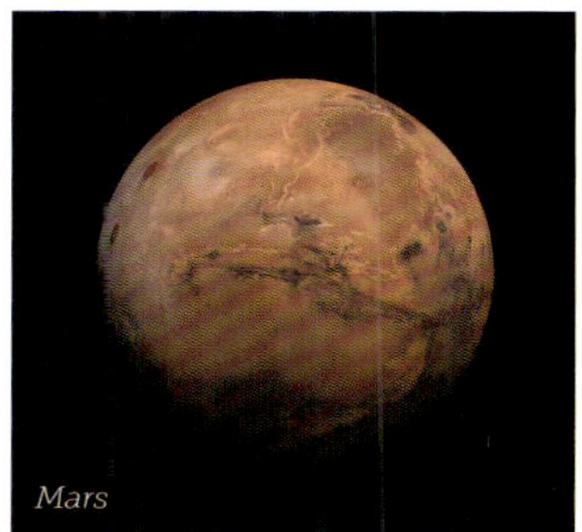

Mars

- This month is your best chance this year to see elusive **Mercury** in the evening sky. It never strays far from the Sun, and even the great astronomer Nicolaus Copernicus – who posited that the planets orbit the Sun – was not able to observe this planet himself.
- You can first catch the innermost planet low in the dusk twilight to the north-west around 8 May, when it's shining at magnitude –1.3 and setting about 9.30 pm. As the days pass, Mercury rises higher above the horizon but it's also fading: your best chance to view it is around 20 May (Chart 5b). By the time it reaches its greatest separation on 28 May, the planet is setting as late as 11 pm, but it has dimmed to magnitude +0.5.
- **Venus**, **Saturn**, **Uranus** and **Neptune** are lost in the Sun's glare this month.

- The sky at 11 pm in mid-June, with Moon positions at three-day intervals either side of Full Moon.
- The star positions are also correct for midnight at the beginning of June, and 10 pm at the end of the month.
- The planets move slightly relative to the stars during the month.

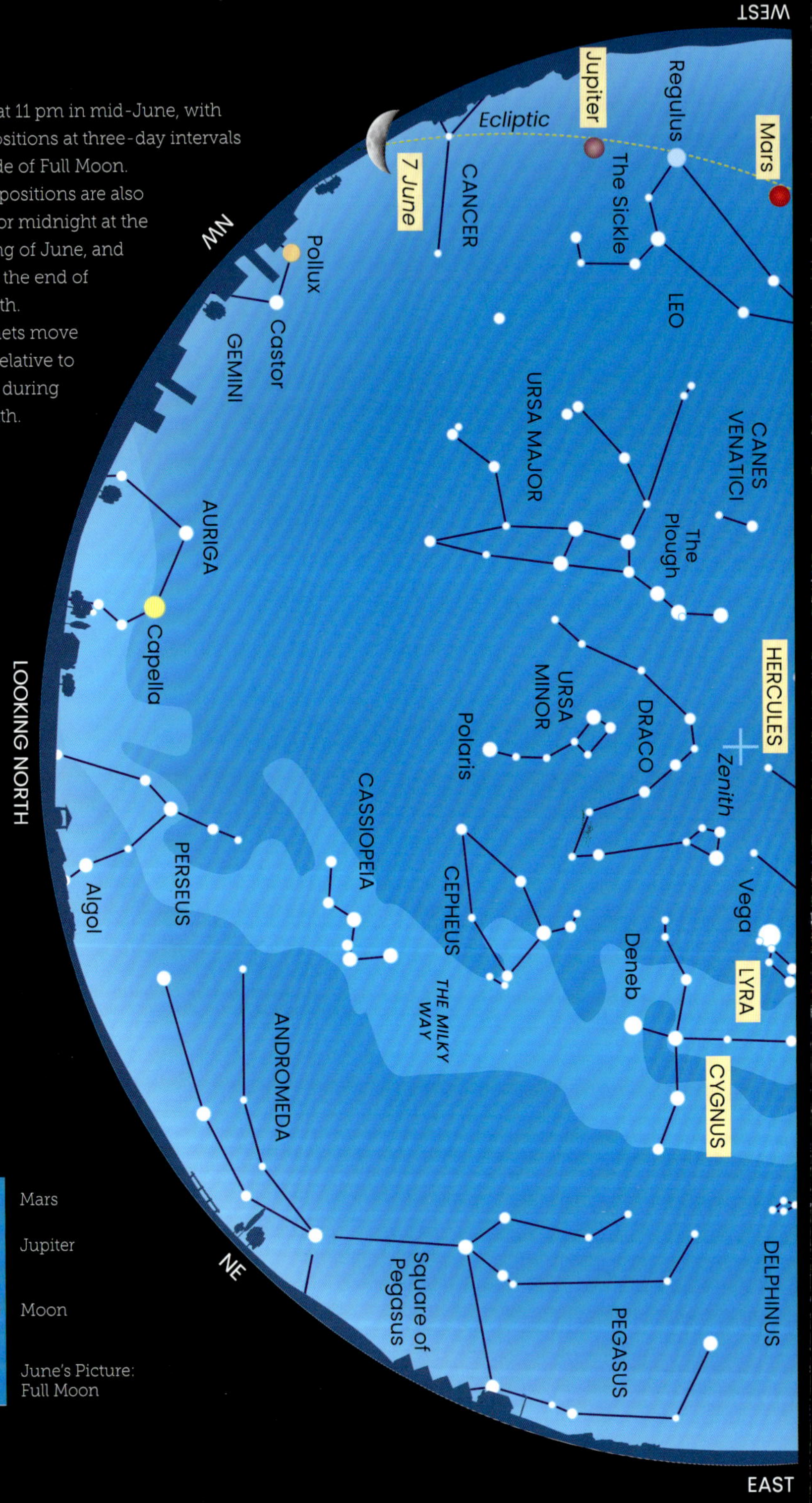

Mars
Jupiter
Moon
June's Picture: Full Moon

JUNE

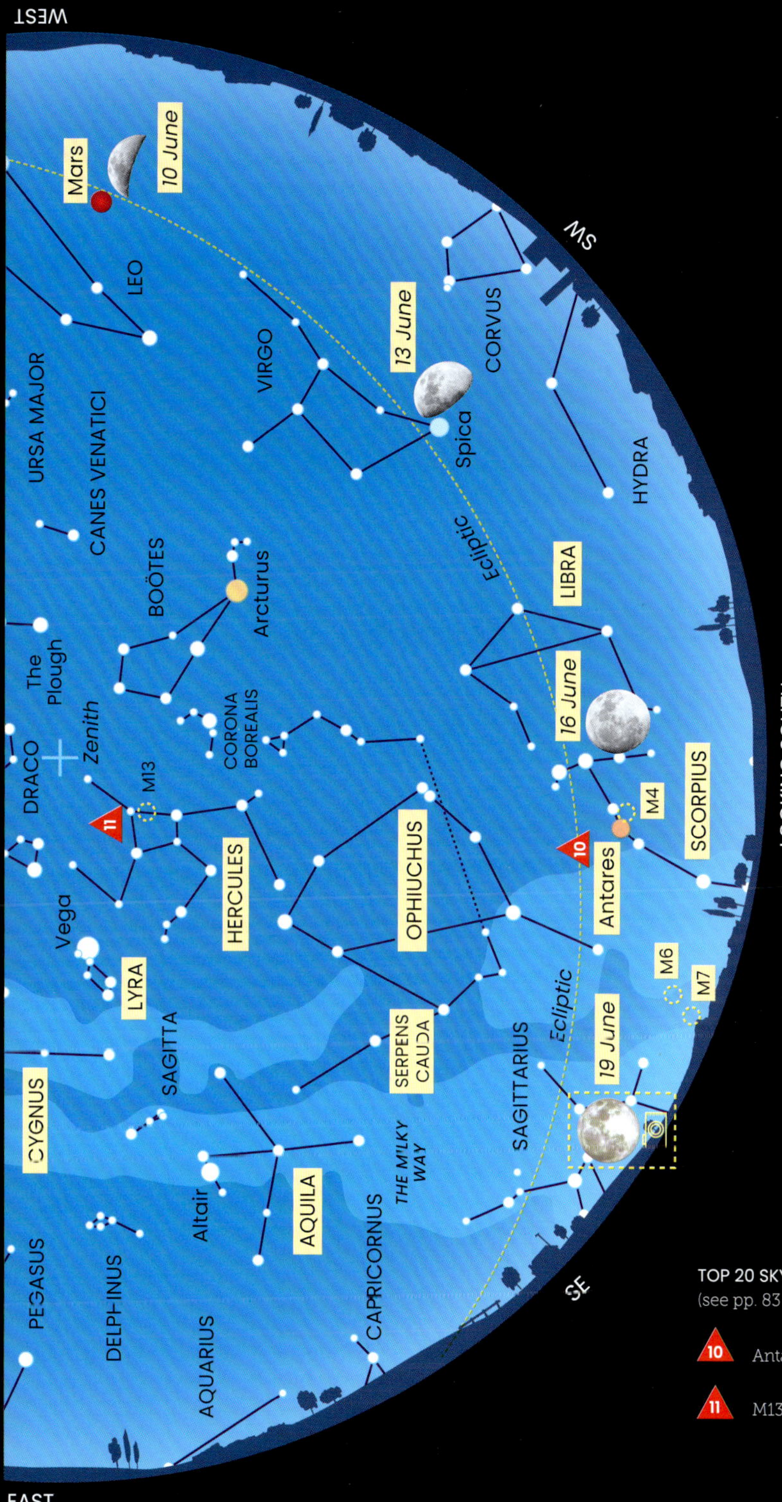

TOP 20 SKY SIGHTS
(see pp. 83–85)

10 Antares

11 M13

JUNE

We still have **Jupiter** and **Mars** adorning Leo to the west, while Saturn reappears in the morning sky. And take advantage of these warmer nights to become acquainted with the lovely summer constellations of **Lyra**, **Cygnus** and **Aquila**, along with the sprawling giants **Hercules** and **Ophiuchus**.

JUNE'S CONSTELLATION

Down in the deep south lies a baleful red star. To the ancient Greeks **Antares** – 'the rival of Mars' – marked the heart of **Scorpius**, the scorpion that killed the great hunter Orion (see January's Constellation).

Scorpius is one of the few constellations that resembles its namesake. A line of stars at the top right marks the scorpion's forelimbs. Originally, the stars we now call **Libra** (the Scales) were its claws. Below Antares, the scorpion's body stretches down into a fine curved tail (below the horizon on the Chart) and deadly sting.

The lovely double stars in Sirius include Antares, its faint companion appearing greenish in contrast to Antares's strong red hue. Binoculars reveal the fuzzy patch of **M4**, a globular cluster made of tens of thousands of stars; at a distance of 6000 light years, it's our nearest globular cluster. Above the 'sting' lie two fine star clusters – **M6** and **M7** – visible to the naked eye when they're well above the horizon: a telescope reveals their stars clearly.

OBSERVING TIP

Don't think that you need a telescope to bring the heavens closer. Binoculars are excellent – and you can fling them into the back of the car at the last minute. For astronomy, buy binoculars with large lenses coupled with a modest magnification. An ideal size is 7 x 50, meaning that the magnification is seven times, and that the diameter of the lenses is 50 millimetres.

JUNE'S OBJECT

Astronomy is such a night-time subject it's easy to overlook the nearest and brightest star, our own **Sun**. Some 150 million kilometres from us, it's 267,000 times closer than Proxima Centauri. And it's the only star we can study in detail.

At its core, where temperatures reach 15.7 million degrees, this giant ball of gas transmutes hydrogen into helium. Every second, the Sun devours 4 million tonnes of itself, converting matter into energy. This intense heat drives currents of incandescent gas up to the solar surface, where it cools and descends again – like boiling water in a pan. The Sun's surface shines at 5500°C. But the thin atmosphere above – the corona – is far hotter, at over a million degrees.

Powerful magnetic fields lace the Sun's surface and corona, creating dark sunspots and giant prominences reaching far into space. Solar magnetism reached the peak of its 11-year cycle in 2024, and solar activity is now fading away towards the next minimum around 2030.

JUNE'S TOPIC: NOCTILUCENT CLOUDS

Look north at twilight, and you may be lucky enough to see what has to be the most ghostly apparition in the night sky

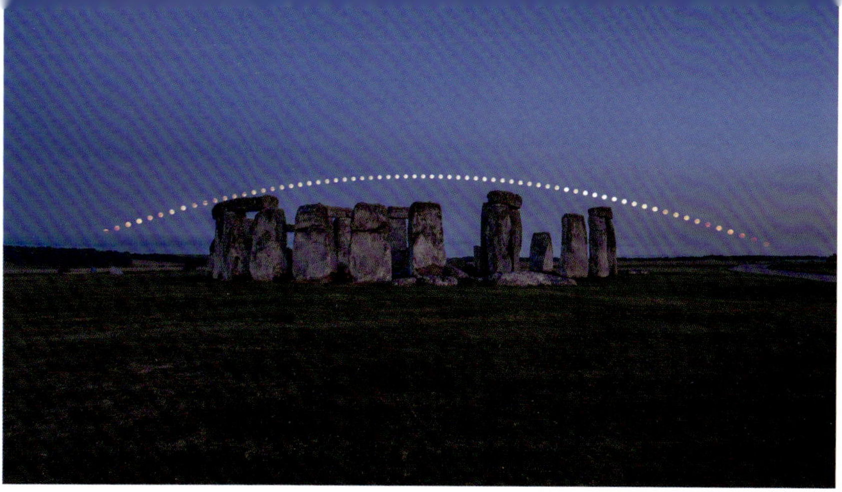

Two years' planning gave Abby Moule the perfect perspective on the major standstill. She carefully calculated her position to ensure that the full arc would fall within the field of view of the 14-mm Sigma Art lens on her Sony A7III camera, 'and that the Moon's transit would just graze the stones.' The image was created during 8 hours of shooting from moonrise to moonset on 10 June 2025, with Abby bracketing shots every 5 minutes. She set the camera to f/4 and ISO 800, with the exposures ranging from 1/60 second at moonrise and moonset, to 1/1000 second when the Moon was at its highest and brightest. She then blended the frames together in Photoshop, along with a long-exposure foreground frame.

– noctilucent clouds. Derived from the Latin for 'night shining', these spooky clouds glow blue-white. Illuminated by the Sun after it's set, they're most commonly seen between latitudes 50 degrees and 70 degrees in the summer, when the Sun doesn't sink far below the horizon even at midnight.

These are the highest clouds in the sky, floating around 80 kilometres up in the atmosphere. And they're composed of ice, coating tiny particles of dust. Some of these dust particles may come from volcanoes: tellingly, the first observation of noctilucent clouds was made in 1885, two years after the great eruption of the Indonesian volcano Krakatoa. But most of the dust particles are probably micrometeorites, raining down from space.

JUNE'S PICTURE

The Full Moon in June is always low down because it lies opposite to the Sun – which is at its highest this month. But in June 2025 the Moon was exceptional, as you can see in Abby Moule's sequence of shots where the Moon skims low over the uprights of Stonehenge as it makes its shortest traverse from moonrise to moonset.

That year marked a 'major lunar standstill', when the Moon was as far below the Sun's path in the sky (the ecliptic) as it can ever get. Some researchers believe that the four Station Stones – outliers at Stonehenge – were set up in line with the Moon's rising and setting at a major lunar standstill.

The Moon's orbit swings around gradually, and the June Full Moon is now higher than it was in 2025. By 2034, the Midsummer Moon will fly well above Stonehenge. For a repeat of the 2025 sight, we'll have to wait until 2042 – and keep our fingers crossed for clear skies!

JUNE'S CALENDAR

SUNDAY	MONDAY	TUESDAY	WEDNESDAY	THURSDAY	FRIDAY	SATURDAY
		1 Moon near Saturn (am)	2	3	4 8.40 pm New Moon	5
6	7	8 Moon near Jupiter	9 Moon between Jupiter and Mars	10 Moon near Mars	11 11.56 am First Quarter Moon	12
13 Moon near Spica	14	15	16 Moon near Antares	17 Moon near Antares	18	19 1.44 am Full Moon
20	21 Summer Solstice	22	23	24	25	26
27 5.54 am Last Quarter Moon	28	29 Moon near Saturn (am)	30			

SPECIAL EVENTS

- **1 June, before dawn:** the crescent Moon hangs above Saturn low in the eastern dawn sky.
- **8 June:** the brilliant 'star' near the Moon is Jupiter, with Regulus and Mars to the left (Chart 6a).
- **9 June:** the Moon lies just below Regulus, with Jupiter to the right and Mars to its left (Chart 6a).
- **10 June:** the Moon passes below Mars, with Regulus and Jupiter to the right (Chart 6a).
- **21 June, 3.11 pm:** Summer Solstice. The Sun reaches its most northerly point in the sky, so today is Midsummer's Day, with the longest period of daylight and the shortest night.

29 June, before dawn: you'll find Saturn below the crescent Moon (Chart 6b).

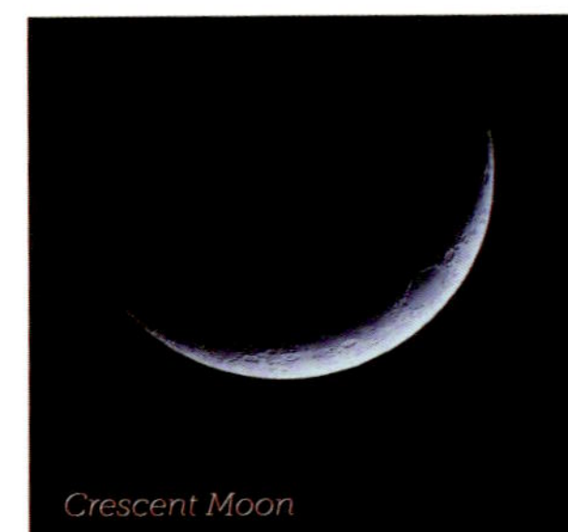
Crescent Moon

JUNE'S PLANET WATCH

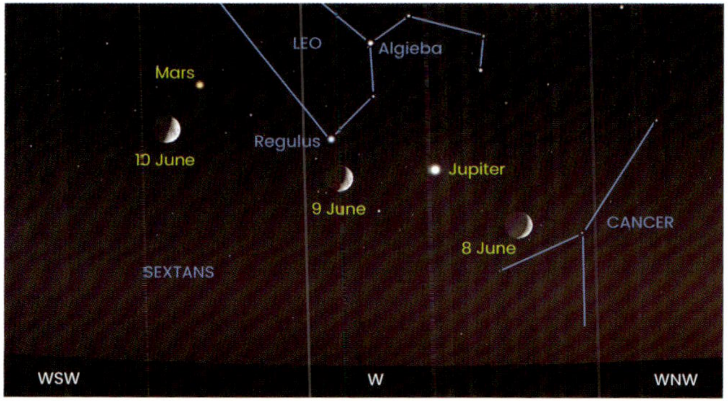

6a *8–10 June, 11 pm. The Moon with Jupiter, Mars and Regulus.*

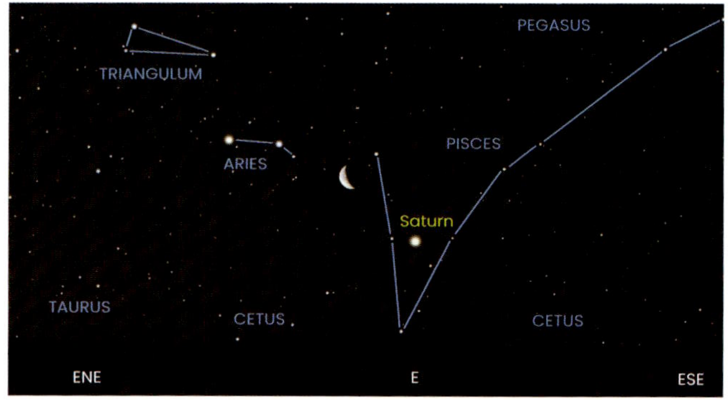

6b *29 June, 2.30 am. The Moon with Saturn.*

- Brilliant **Jupiter** lords it over the western sky until the giant planet sets around 0.30 am. Shining more brightly than any of the stars – at magnitude –1.9 – Jupiter lies in Leo, to the right of Regulus. The Moon passes nearby on 8 and 9 June (Chart 6a).

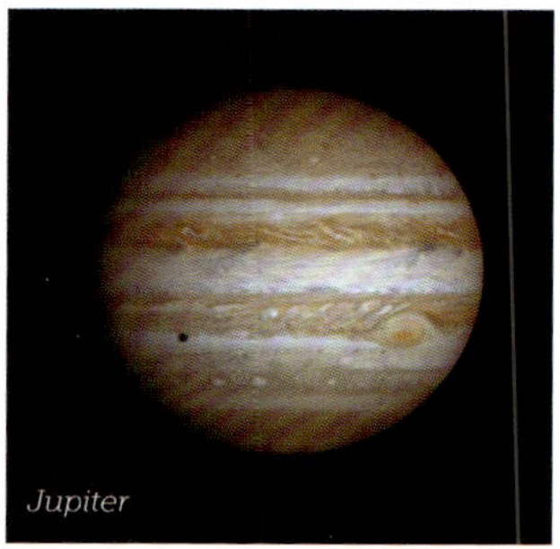

- You'll find **Mars** below the main stars of Leo, at magnitude +0.8, and setting about 1 am. During June, the Red Planet heads rapidly to the left, towards the constellation Virgo. The Moon sails underneath Mars on 10 June (Chart 6a).
- With a telescope, you can catch **Neptune** (magnitude +7.8) rising around 2 am in Pisces.
- **Saturn** is now re-emerging in the dawn sky. It shines at magnitude +0.7, also in Pisces, and rises about 2.30 am. The crescent Moon hangs directly over the ringworld on the morning of 1 June, and revisits it on 29 June (Chart 6b).
- **Mercury**, **Venus** and **Uranus** are too close to the Sun to be seen in June.

- The sky at 11 pm in mid-July, with Moon positions at three-day intervals either side of Full Moon.
- The star positions are also correct for midnight at the beginning of July, and 10 pm at the end of the month.
- The planets move slightly relative to the stars during the month.

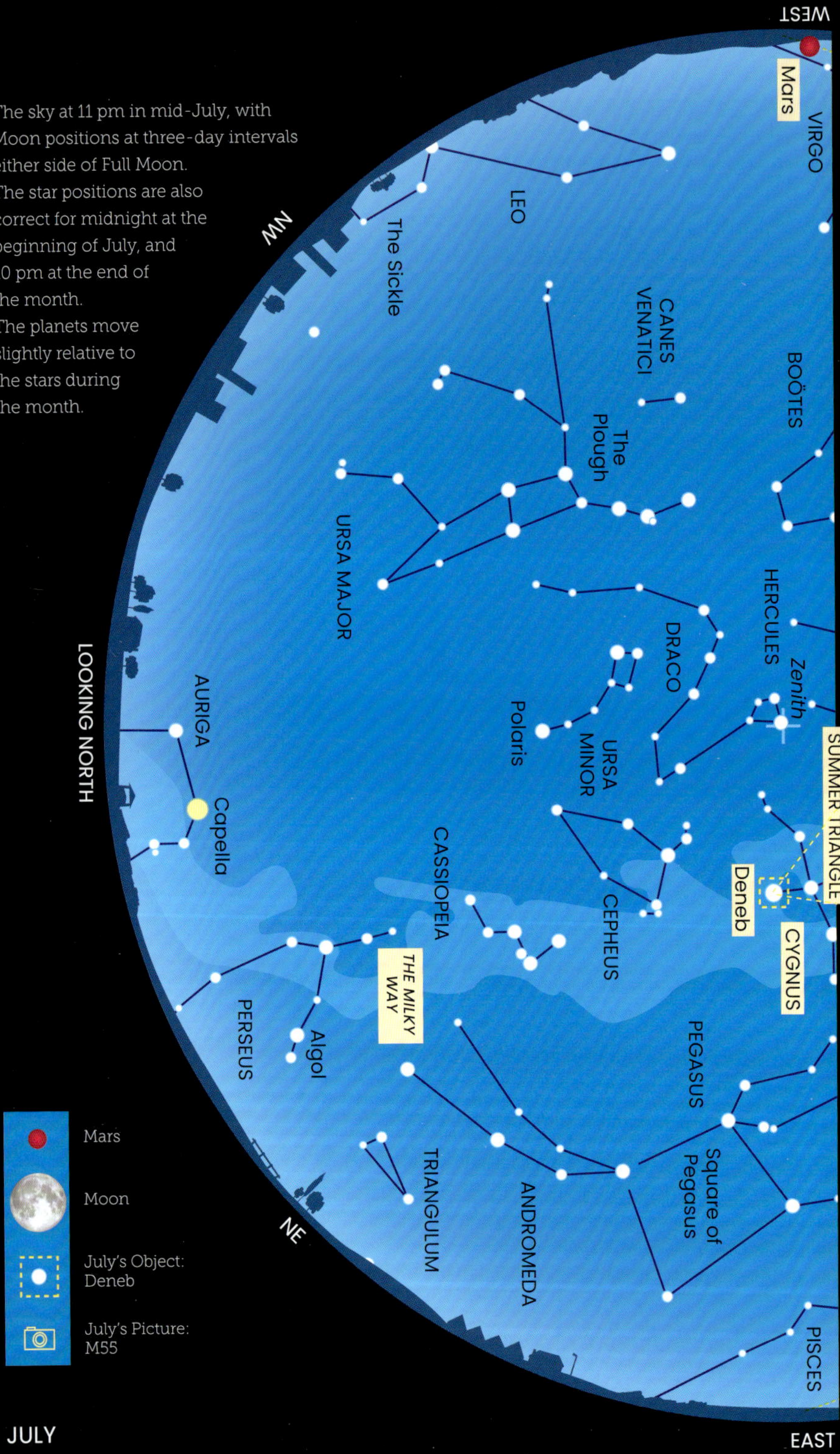

JULY

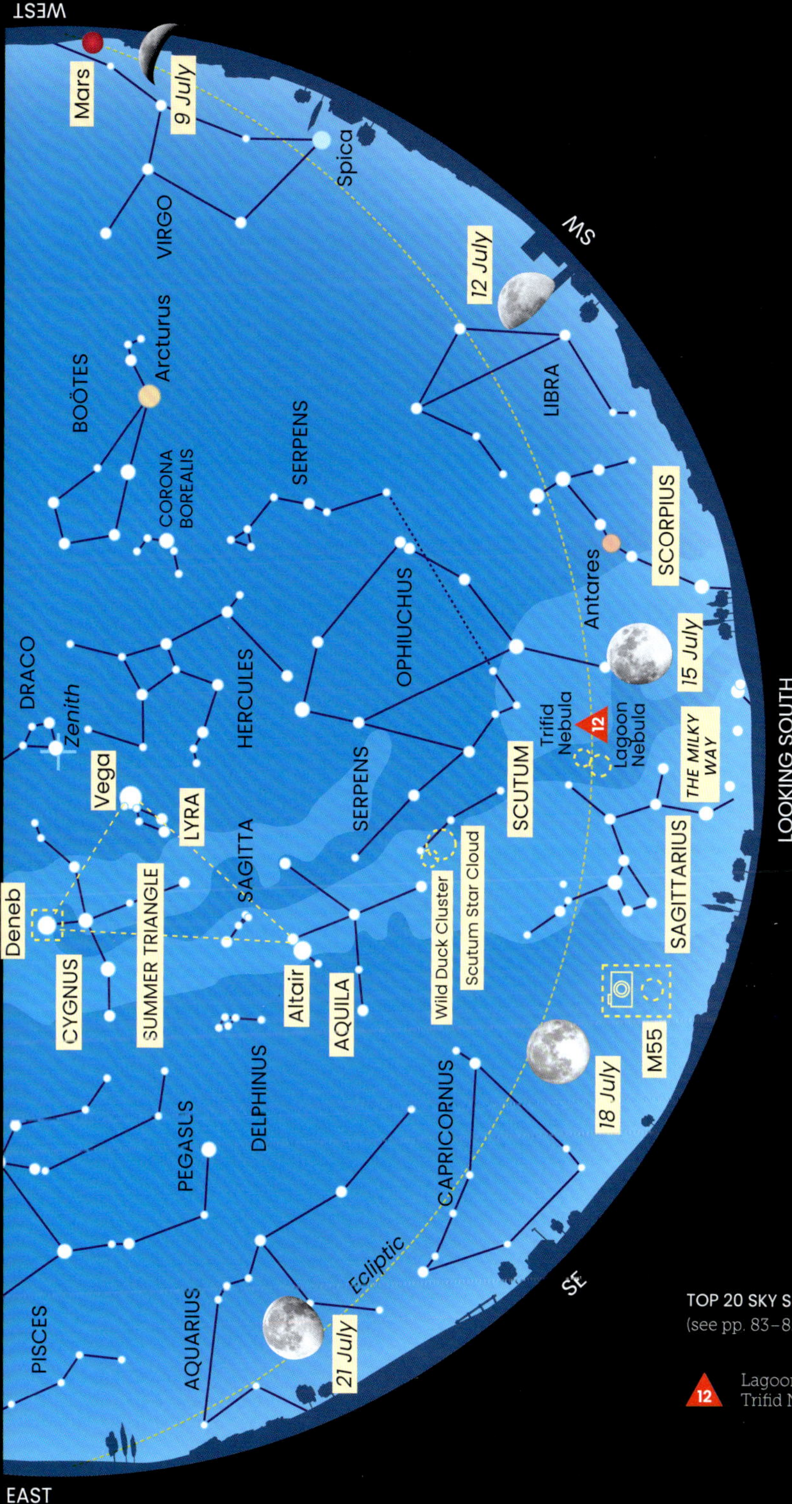

TOP 20 SKY SIGHTS
(see pp. 83–85)

12 Lagoon and Trifid Nebulae

Two of the most ancient constellations, **Sagittarius** and **Scorpius**, are at their best this month decorating the southern region of the sky. Higher up, three brilliant stars – **Vega**, **Deneb** and **Altair** – mark out the vertices of the **Summer Triangle**.

JULY'S CONSTELLATION

One of the 'new' constellations – meaning it doesn't date back to the ancient Greeks – **Scutum** (the Shield) was created by Polish astronomer Johannes Hevelius in 1684. Originally named Scutum Sobiescianum, it commemorated the shield borne by his victorious king, Jan Sobieski, at the Battle of Vienna.

You can locate Scutum most easily not by its stars (which are all below magnitude +3.8), but by a small bright patch of the **Milky Way**, known as the **Scutum Star Cloud**. Framed by surrounding dark nebulosity, this glowing region is a window to distant regions of the Milky Way. Near the top of this cosmic peephole, some 6000 light years away, lies the **Wild Duck Cluster** (M11), a compact swarm of a thousand stars. It's a lovely sight in a small telescope: see if you agree with the Victorian astronomer William Henry Smyth who thought it resembled 'a flight of wild ducks'.

OBSERVING TIP

This is the month when you really need a good, unobstructed horizon to the south for the best views of the glorious summer constellations of Scorpius and Sagittarius. They never rise high in temperate latitudes, so make the best of a southerly view – especially over the sea – if you're away on holiday. A good southern horizon is also best for views of the planets, because they rise highest when they're in the south.

JULY'S OBJECT

Deneb – marking the tail of the celestial swan, **Cygnus** – may look the least significant of the three stars marking the vertices of the **Summer Triangle** – but appearances are deceptive. In reality, Deneb is over a thousand times more luminous than either of the other two Summer Triangle stars, **Vega** and **Altair**. It only appears slightly fainter because of its immense distance – so far off that astronomers can't get an exact handle on Deneb's remoteness, with estimates ranging from 1400 to 2600 light years. Either way, it's the most distant of the brightest 30 stars in the sky.

A heavyweight star, some 19 times heavier than the Sun, Deneb is white-hot, with surface temperature of 8400°C. It has swollen to become a supergiant as it approaches the end of its life: placed in the centre of the Solar System, it would almost fill the Earth's orbit. Within a few million years – an eyeblink on the astronomical scale – Deneb will blow itself apart in a supernova explosion.

JULY'S TOPIC: GALAXIES

The glowing band of the Milky Way – stretching from the southern horizon up through the sky – is the inside view of our Galaxy, a cosmic metropolis of 400 billion stars, of which our Sun is an insignificant member. From within, it's difficult to work out the shape of our Galaxy, but looking far out into space we can see that galaxies come in three basic types.

'Elliptical galaxies' are simple balls of elderly orange and red stars, without much trace of gas and dust that could condense into new stars. They range in size from some of the smallest known galaxies to the biggest galaxies in the Universe.

'Irregular galaxies' are tiny, and – as their name suggests – merely ragged bundles of stars and gas lacking any particular shape. They are the survivors of the basic building blocks of the Universe: most of the original irregular galaxies merged to form the larger star cities around us today.

And that brings us to the third – and most spectacular – type. 'Spiral galaxies' have a central bulge of old stars that sprouts two or more beautiful curving arms, where gas and dust are condensing into new generations of stars. Careful mapping of the Milky Way shows that we live in a spiral galaxy – it's a pity we can never appreciate its stunning shape from the outside!

JULY'S PICTURE

A ball of 200,000 stars low down in **Sagittarius**, **M55** was discovered by French astronomer Nicolas-Louis de Lacaille observing near the Cape of Good Hope, southern Africa, in 1752. He described it as resembling 'an obscure nucleus of a big comet'.

Even a small telescope will resolve M55 into a multitude of colourful stars, well on display in Damian Peach's superb image. This is an old cluster, born soon after the Big Bang, and many of its stars have cooled and expanded to become red giants. Scattered between these ruby gems are pale sapphires, stars that have evolved beyond the red giant stage and shine with a blue-white heat as they burn helium to carbon in their cores.

M55 is one of about 160 globular clusters that swarm around the Milky Way galaxy, and – like many of them – it lies towards the Galaxy's core. Some 17,000 light years from us, the cluster is only 10,000 light years from the galactic centre.

Damian Peach captured M55 with a 500-mm Corrected Dall-Kirkham reflector, which gives pin-sharp star images over a wide field of view. He used an FLI ProLine 16803 CCD camera, and took a total of 2 hours' exposure through luminance (white light), red, green and blue filters.

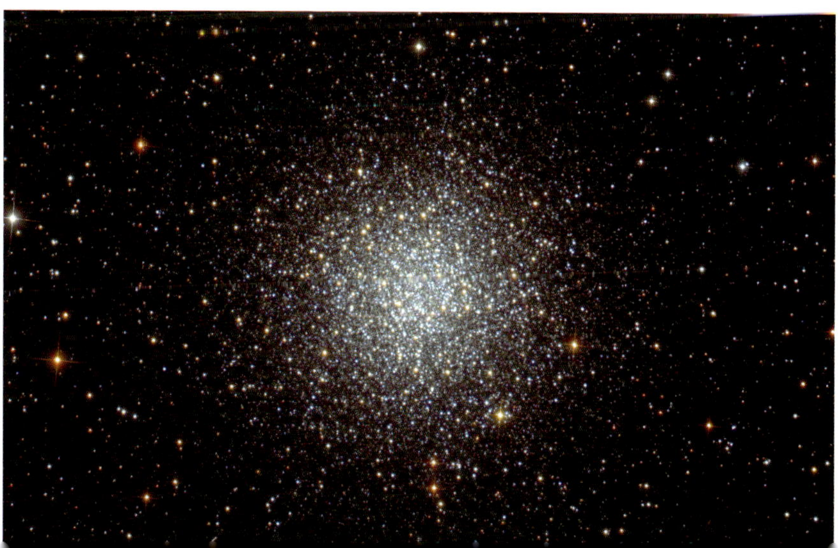

JULY'S CALENDAR

SUNDAY	MONDAY	TUESDAY	WEDNESDAY	THURSDAY	FRIDAY	SATURDAY
				1	2	3
4 4.02 am New Moon	5 Earth at aphelion	6 Moon near Jupiter	7 Moon near Jupiter	8 Moon near Mars	9	10 7.39 pm First Quarter Moon near Spica
11	12	13	14 Moon very near Antares	15 Mercury W elongation	16	17
18 4.45 pm Full Moon	19	20	21	22	23	24
25	26 5.55 pm Last Quarter Moon near Saturn (am)	27	28	29	30	31

SPECIAL EVENTS

- **5 July, 6.06 am:** the Earth is furthest from the Sun (aphelion), at 152 million km.
- **6 July:** a lovely sight in the evening twilight, as a narrow crescent Moon lies just below Jupiter, low in the west (Chart 7a).
- **7 July:** the dusk sky is enlivened by a crescent Moon lying to the left of Jupiter (Chart 7a).
- **8 July:** the reddish 'star' that can be observed close to the Moon is the planet Mars (Chart 7a).
- **14 July:** the Moon skims just below the bright star Antares; as seen from southern Africa, the red giant is occulted by the Moon.
- **15 July:** Mercury reaches its greatest morning separation from the Sun (see Planet Watch).
- **26 July, before dawn:** you'll find Saturn below the Last Quarter Moon.

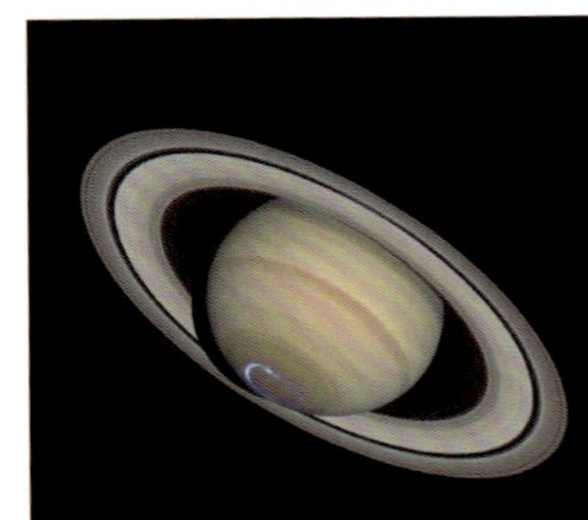

Saturn

JULY'S PLANET WATCH

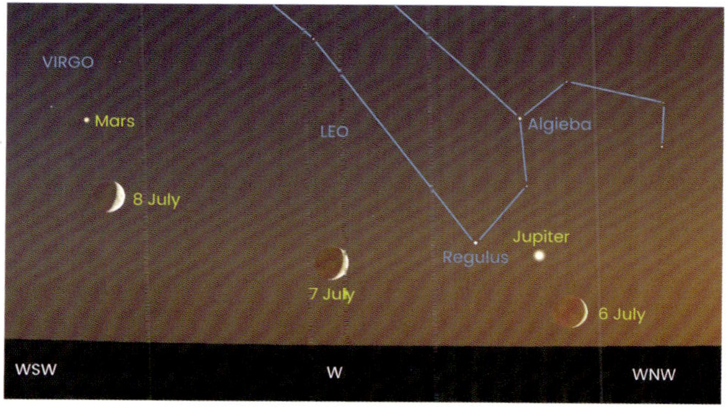

7a *6–8 July, 10.30 pm. The Moon passes Jupiter and Mars.*

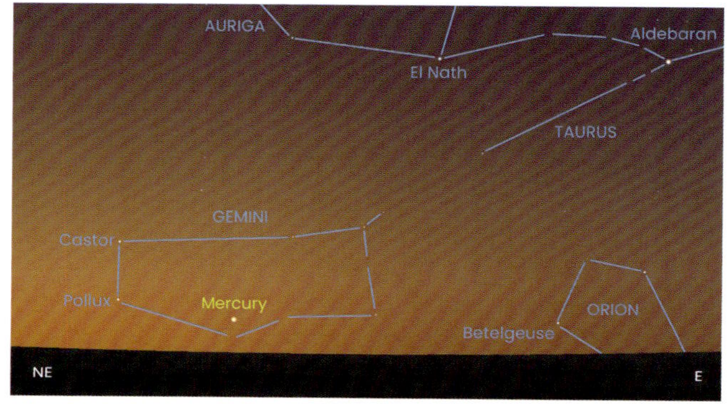

7b *27 July, 4.15 am. Mercury in the dawn sky.*

- Brilliant **Jupiter** is low in the western sky after sunset, shining at magnitude –1.8 in Leo, just to the right of Regulus. The crescent Moon passes by the planet on 6 and 7 July (Chart 7a). Jupiter is setting about 10.30 pm, and by the end of the month it has disappeared into the twilight glow.
- **Mars**, at magnitude +1.1, is in Virgo and sets at about 11.30 pm. The Moon lies nearby on 8 July (Chart 7a).
- Rising around 0.30 am, **Saturn** shines at magnitude +0.6 in Pisces. The Moon passes over Saturn on the morning of 26 July.
- Faint **Neptune** (magnitude +7.8) lies to the upper right of Saturn, also in Pisces, and clears the horizon about midnight.
- At magnitude +5.8, **Uranus** lies in Taurus and rises around 2 am.
- **Mercury** reaches greatest western elongation on 15 July, and for the second half of the month you can catch it very low in the north-east where it rises about 3.45 am (Chart 7b). Between 15 and 31 July, the innermost planet brightens from magnitude +0.3 to magnitude –1.3.

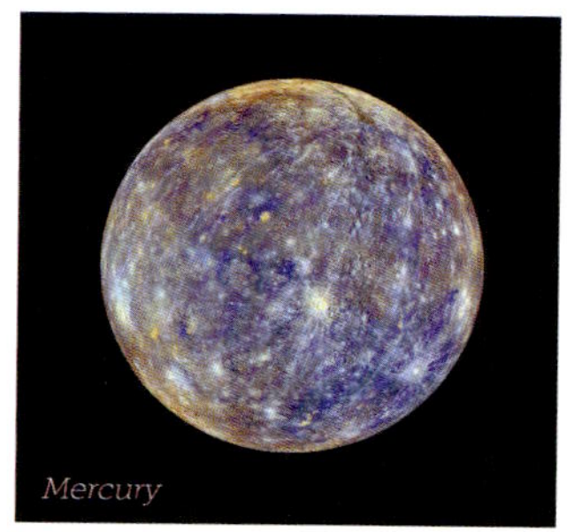

Mercury

- **Venus** is lost in the Sun's glare this month.

- The sky at 11 pm in mid-August, with Moon positions at three-day intervals either side of Full Moon.
- The star positions are also correct for midnight at the beginning of August, and 10 pm at the end of the month.
- The planets move slightly relative to the stars during the month.

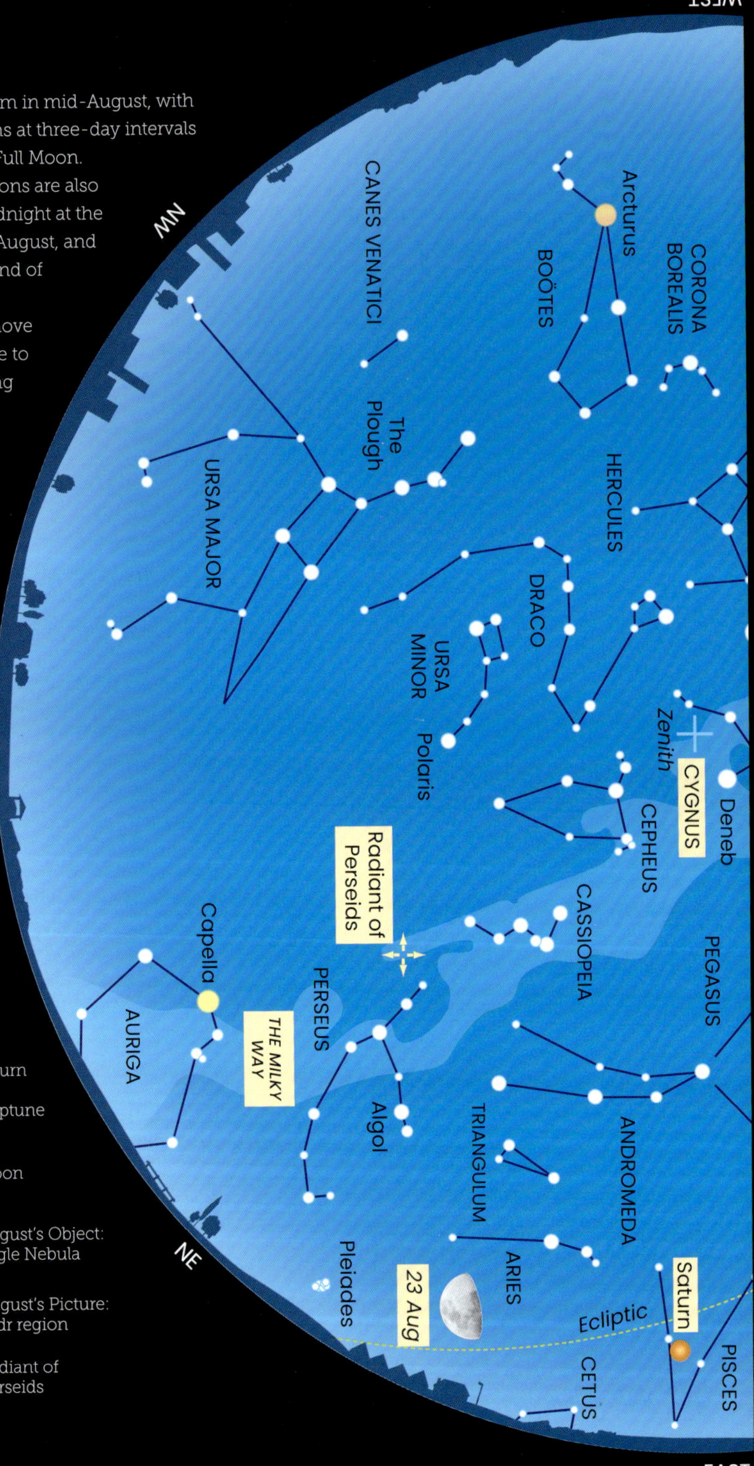

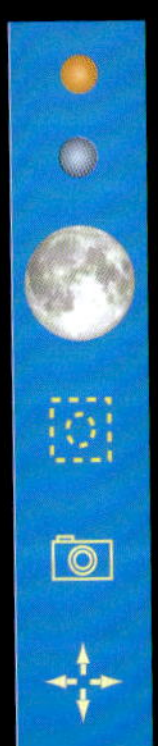

AUGUST

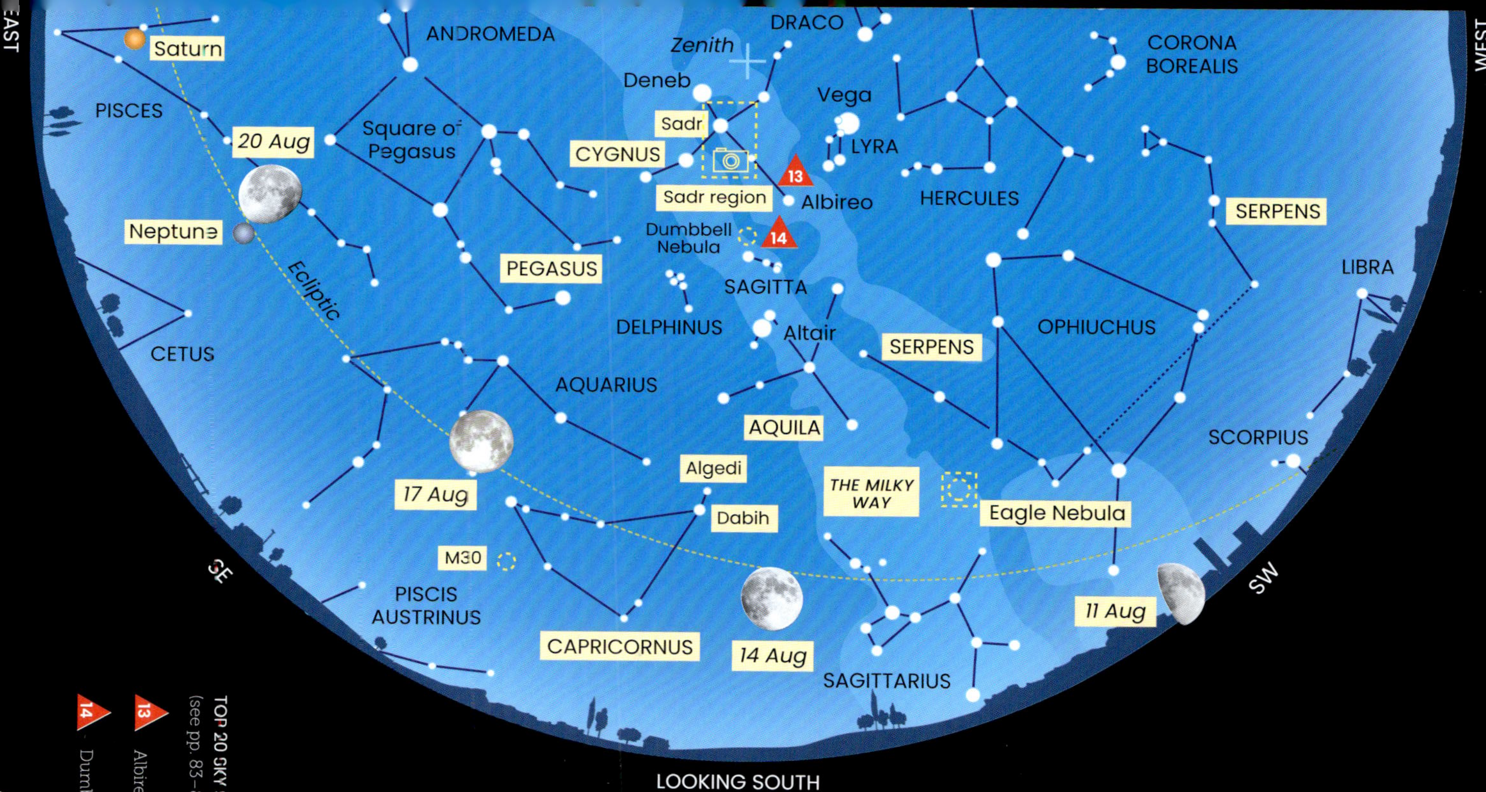

TOP 20 SKY SIGHTS
(see pp. 83–85)

- 13 Albireo
- 14 Dumbbell Nebula

We're treated to a spectacular solar eclipse this month: travel to northern Africa if you can for the unforgettable sight of a total eclipse; follow the event live on social media; or observe a large partial eclipse from your own backyard in the UK or Ireland.

AUGUST'S CONSTELLATION

Capricornus (the Sea Goat) is one of the faint watery constellations below **Pegasus**, its triangular outline looking like a boat. The ancient Sumerians associated these stars with their great deity, Enki, who presided over both the sea and the creation of life – and so was represented by a strange being who was half-fish and half-goat.

Even with the unaided eye, you can see that **Algedi** is double. The two stars (magnitudes +3.6 and +4.3) aren't actually related, though, but happen to lie one behind the other.

The next-door star **Dabih** is a genuine binary. The brighter member is a yellow star of magnitude +3.1; binoculars or a small telescope will reveal a blue companion at magnitude +6.1.

Using a Samyang 135-mm lens set at f/2.8, Tracey Harty captured these glowing gas clouds with a cooled one-shot colour ZWO ASI071 camera. She took 40 × 4-minute exposures, for a total integration time of 2 hours 40 minutes, and processed them in PixInsight and Adobe Lightroom.

You can spot the globular cluster **M30** in binoculars, and a telescope reveals its swarming stars. About 27,000 light years away, this eighth-magnitude star cluster, orbits the opposite way to our Galaxy's rotation, so the smart money says the Milky Way purloined M30 from a passing galaxy.

AUGUST'S OBJECT

The **Eagle Nebula** – no relation to the constellation **Aquila**! – is the go-to destination in the constellation of **Serpens** (the Serpent). In short exposures, its shape resembles a flying bird. Deeper photographs reveal three dark fingers of cosmic dust, where new stars and planetary systems are being hatched, silhouetted against the glowing gas clouds. These Pillars of Creation became a celestial legend after they appeared in a stunning image from the Hubble Space Telescope. The James Webb Space Telescope revealed more intimate details, including newly born stars within the Pillars.

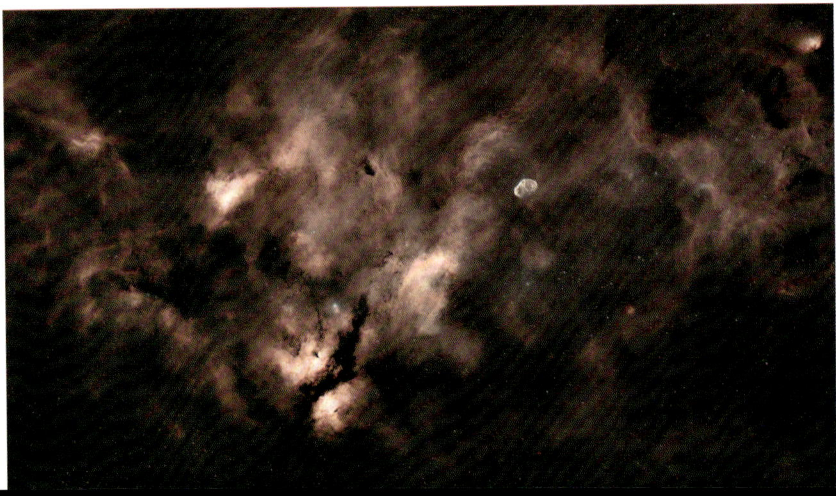

Catalogued as M16, the Eagle Nebula was discovered by Jean-Philippe Loys de Cheseaux in 1745. The gas cloud is 5700 light years away, and it's home to thousands of stars that are less than 2 million years old (less than a thousandth the age of our Sun). Through binoculars, you'll easily spot the bright baby stars, but you'll need a telescope to view the nebulosity.

AUGUST'S TOPIC: TOTAL SOLAR ECLIPSE

It's being billed as the 'Eclipse of the Century'! As the Moon's shadow sweeps along a thin track on the Earth, from the Straits of Gibraltar to the Horn of Africa, some 88 million people will be treated to nature's most awesome spectacle, a total eclipse of the Sun. And over half the world's population will see a partial eclipse.

The date is 2 August, and the total eclipse begins for southern Spain, Gibraltar and Morocco in the morning. Next up are northern Algeria, Tunisia, Libya and Egypt. You'll see the maximum duration – 6 minutes 23 seconds – near the Egyptian city of Luxor, where the eclipsed Sun will be directly over the ancient temples. The path of totality travels down the Red Sea and ends in the Indian Ocean.

The eclipse is partial over all of Europe, the Middle East, India and most of Africa. From the UK, about 40 per cent of the Sun is covered (see Special Events), and the maximum eclipse is around 10 am (the exact time depends on your location).

During the partial phase, make sure your eyes are protected (see Observing Tip). During the total eclipse, you can look at the Sun directly to witness its glowing corona – like a luminous flower with a dark heart – before totality ends in a brilliant diamond ring.

OBSERVING TIP

Whether you're observing the partial solar eclipse this month, or just checking out our local star, NEVER look at the Sun directly with your unprotected eyes or – especially – with a telescope or binoculars: it can blind you permanently. For naked-eye observing, use special 'eclipse glasses' with dark filters (meeting the ISO 12312-2 safety standard). Check out the Observing the Sun chapter (pages 86–89) for info on how to observe safely with a telescope or take photos.

AUGUST'S PICTURE

If our eyes could easily see the red light emitted by hot hydrogen gas, the central part of Cygnus – the **Sadr region** – would be the most stunning sight in the northern sky, as revealed in Tracey Harty's haunting image.

The central star is **Sadr**, the middle star of **Cygnus**, and the view stretches across 10 degrees of sky (20 times the Full Moon's width). Sadr itself is in the foreground, with the glowing gas clouds lying some 5000 light years away.

The small oval (centre right) is the Crescent Nebula, a ring of glowing gas shrugged off by an intensely hot Wolf-Rayet star near the end of its life. At the top right lies the Tulip Nebula, illuminated by an incandescent young star.

But what to make of the great tangle of glowing gas clouds filling the centre of this view? In 1959, American astronomer Stewart Sharpless listed all the nebulae (technically, 'HII regions') in the northern Milky Way – but the structure of this one baffled him. He noted 'the great size and gross filamentary structure of this nebula set it apart from any known HII region'.

AUGUST'S CALENDAR

SUNDAY	MONDAY	TUESDAY	WEDNESDAY	THURSDAY	FRIDAY	SATURDAY
1	2 11.05 am New Moon; total solar eclipse	3	4	5	6 Moon between Mars and Spica	7 Moon near Spica
8	9 5.54 am First Quarter Moon	10 Moon near Antares	11	12 Perseids	13 Perseids (am)	14
15	16	17 8.29 am Full Moon	18	19	20	21 Moon near Saturn
22 Moon near Saturn	23 Mars near Spica	24 Mars near Spica; Moon occults the Pleiades	25 Moon occults the Pleiades (am); 3.27 am Last Quarter Moon	26	27	28
29	30	31 6.41 pm New Moon				

SPECIAL EVENTS

- **2 August:** a total solar eclipse is visible along a narrow band of north Africa (see Topic). For the UK and Ireland, it's a partial eclipse: from Cornwall you'll see almost 45 per cent of the Sun covered by the Moon (Chart 8a), falling to 20 per cent for the north of Scotland.
- **6 August:** as the sky grows dark, look out for the crescent Moon with Mars to the upper right and Spica to its upper left.
- **Night of 12/13 August:** the maximum of the **Perseid meteor shower** is spoilt by a bright gibbous Moon. After moonset – in the small hours of the morning – you'll have the best views of these speedy dust pellets from Comet Swift–Tuttle burning up above our heads.
- **21 August:** the 'star' to the lower left of the Moon is Saturn.
- **22 August:** Saturn lies to the lower right of the Moon.
- **23 August:** Mars passes less than 2 degrees from Spica.
- **24 August:** Spica is directly below Mars
- **Night of 24/25 August, 10.45 pm–1.30 am:** the Moon moves across the centre of Pleiades, occulting in turn most of its brighter stars (Chart 8b).

AUGUST'S PLANET WATCH

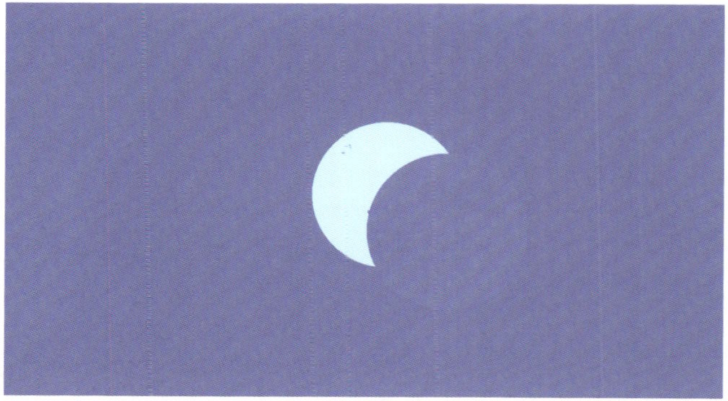

8a *2 August, 9.55 am (view from Penzance, Cornwall). Partial solar eclipse.*

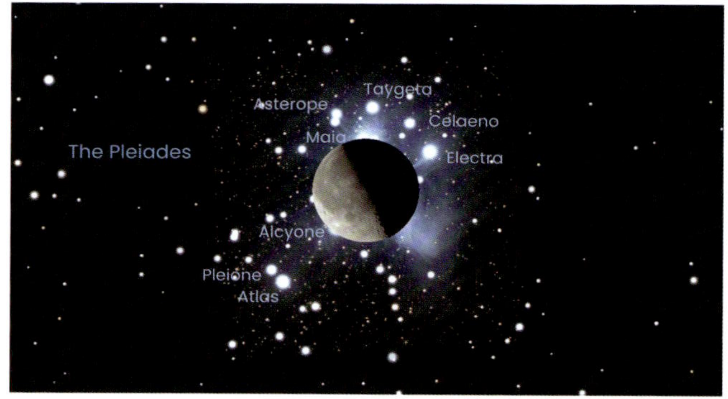

8b *25 August, 0.00 am. The Moon occults the Pleiades.*

- **Mars** is low in the west after sunset, setting around 10 pm. Travelling through Virgo this month, the planet is slightly fainter (magnitude +1.3) than the constellation's principal star, Spica (magnitude +1.0) There's a lovely sight on 6 August when the crescent Moon lies below Mars, with Spica to the left. Mars passes Spica on 23–24 August (see Special Events): binoculars will show the striking colour contrast between the Red Planet and the blue-white star.
- Rising about 10.30 pm, **Saturn** (magnitude +0.4) lies in Pisces. The Moon is above the ringworld on 21 and 22 August.
- **Neptune** – also in Pisces – rises around 9.30 pm and is below naked-eye visibility at magnitude +7.7.
- Neptune's slightly brighter (magnitude +5.8) twin, **Uranus**, lies in the constellation of Taurus and rises about midnight.

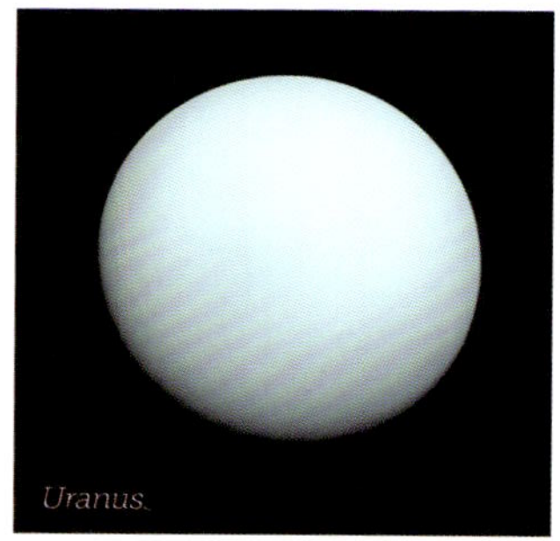

- On the first few mornings of August, you may catch **Mercury** shining at magnitude –1.4 low in the north-east before sunrise, but it then disappears into the dawn glow. The planet reappears deep in the evening twilight right at the end of the month, at magnitude –0.5
- **Venus** and **Jupiter** are too close to the Sun to be visible in August.

- The sky at 11 pm in mid-September, with Moon positions at three-day intervals either side of Full Moon.
- The star positions are also correct for midnight at the beginning of September, and 10 pm at the end of the month.
- The planets move slightly relative to the stars during the month.

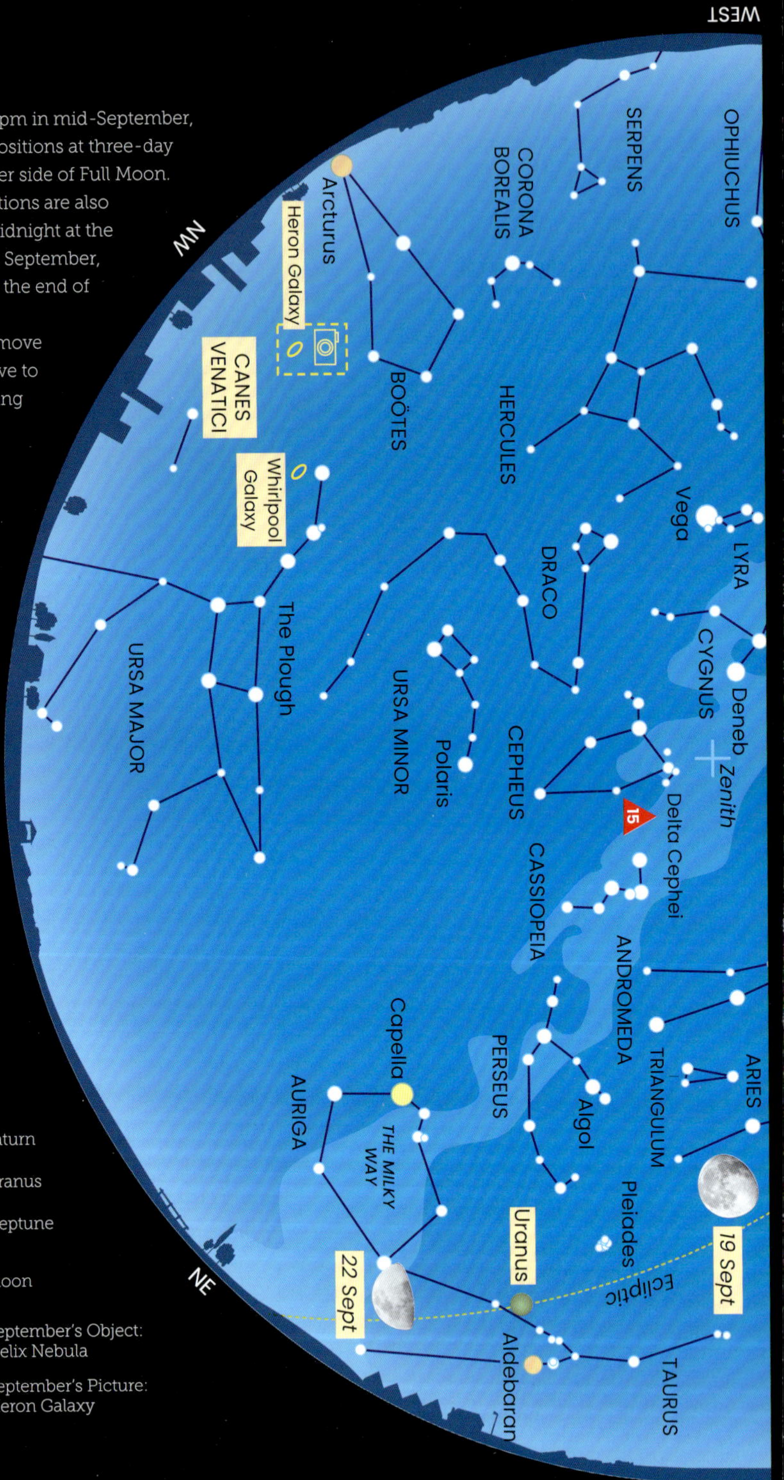

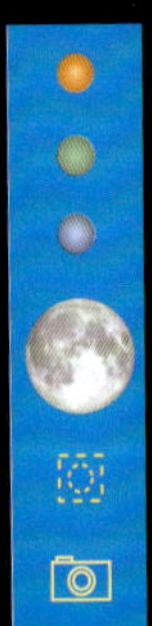

SEPTEMBER

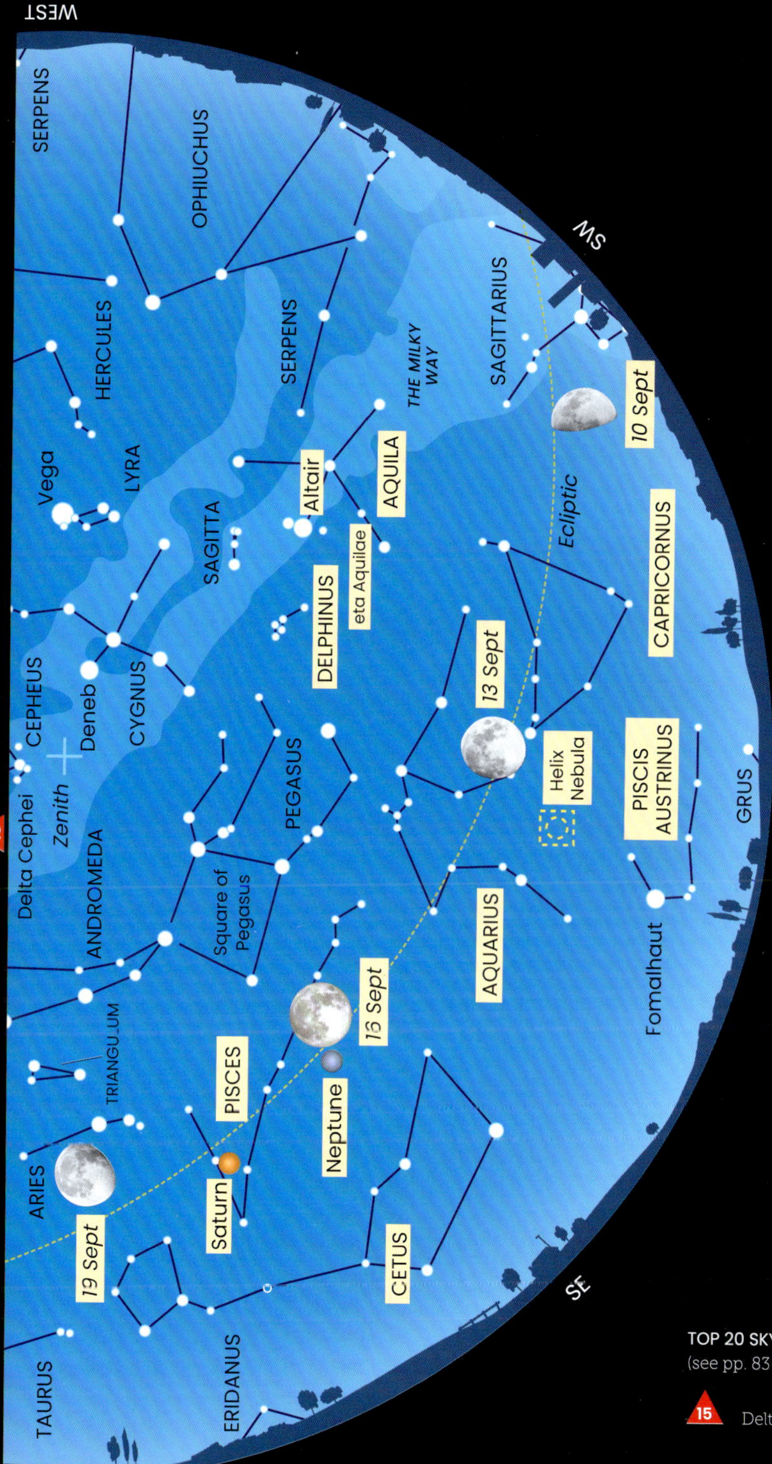

TOP 20 SKY SIGHTS
(see pp. 83–85)

15 Delta Cephei

The autumnal sky is awash with watery constellations: **Piscis Austrinus** (the Southern Fish) and **Aquarius** (the Water Carrier) are joined by **Delphinus** (the Dolphin), the strangely named Sea Goat (**Capricornus**), a pair of Fishes (**Pisces**) and **Cetus** (the Sea Monster).

SEPTEMBER'S CONSTELLATION

Soaring over the September landscape, **Aquila** (the Eagle) is named after the bird that was a companion to Jupiter. The pet eagle carried the mighty god's thunderbolts, and scooped up the Trojan youth Ganymede when Jupiter took a shine to the handsome youth.

The constellation is dominated by **Altair**, a young white star 17 light years away, which is 11 times brighter than our Sun. It has a very fast spin, rotating in just 9 hours (as opposed to about a month for the Sun), so its equator is hurtling around at 280 kilometres per second.

Eta Aquilae is one of the brightest Cepheid variable stars – old stars which change their brightness by swelling and shrinking. The pulsations of eta Aquilae make it vary from magnitude +3.5 to +4.4 every seven days.

SEPTEMBER'S OBJECT

If you'd like a time machine to transport you 6 billion years into the future, look no further than the constellation **Aquarius**. Visible in binoculars or a small telescope, the **Helix Nebula** is a snapshot of the Sun in aeons to come.

The original star's blazing core has collapsed into a superdense white dwarf. But it's overshadowed by the spectacle of the star's outer layers: ejected 10,000 years ago, they have expanded into a glorious oval of glowing gas almost three light years across.

A flood of ultraviolet radiation from the central star, at a temperature of 120,000°C, makes the ejected gas fluoresce spookily, like an ultraviolet 'black light' bulb shining on a white shirt. Rings of denser gas give the Helix Nebula its nickname, while high-resolution images reveal over 40,000 globules of gas, each larger than the Solar System.

This type of stellar wraith is called a planetary nebula – because it resembles a dim planet through a small telescope – and the Helix is particularly important as it's the nearest bright planetary nebula, only 650 light years away.

OBSERVING TIP

When you first go out to observe, you may be disappointed at how few stars you can see in the sky. But wait for 20 minutes, and you'll be amazed at how your night vision improves. One reason for this 'dark adaptation' is that the pupil of your eye grows larger. More importantly, in dark conditions the retina of your eye builds up bigger reserves of rhodopsin, the chemical that responds to light.

SEPTEMBER'S TOPIC: LIBRATION

Night after night, month after month, we always see the same side of the Moon facing us – the smiling visage of the 'Man in the Moon'. It's because our companion world turns on its axis in the same time that it takes to orbit the Earth.

This synchronicity suggests that we can only make out 50 per cent of

An impressively large telescope – a home-built 600-mm f/4 Dobsonian – helped Tom Williams capture the Heron Galaxy in exquisite detail. From his backyard in Wiltshire, he took 7000 × 1-second exposures with a Player-One Apollo-M MAX camera for luminance, plus 11,000 × 1-second exposures with a Uranus-C colour camera, for a total exposure time of 5 hours.

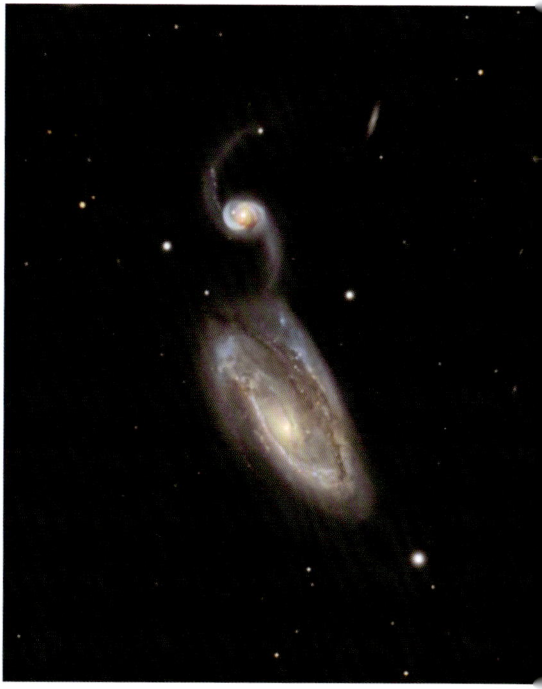

the Moon's surface from the Earth. But around 59 percent of the lunar landscape is actually visible, because at times we can see slightly around the edge of the Moon and catch glimpses of its far side.

This effect is called 'libration', and it arises for two reasons. The Moon's axis is tilted, so sometimes we can see over its north pole and at other times beyond the south pole. Also, the Moon's rotation rate is always constant, but its speed around the Earth varies as it follows an elliptical orbit. When it's travelling slowly, the Moon's rotation brings some of its farside into view around the equator; when it's speedy, we see around the other side.

This month, the two effects combine to show us a huge impact crater on the lunar farside, which appears on the lower left edge of the Moon as we view it from Earth. Long mountain ranges and dark plains – all foreshortened by perspective – are a glimpse of the 900-kilometre-diameter Mare Orientale. For the best view of this lunar farside feature, make an appointment with the Moon at 10 pm on 19 September (Chart 9a).

SEPTEMBER'S PICTURE

Tilt the page a bit clockwise, and you can see why this gorgeous celestial object is nicknamed the **Heron Galaxy**! The bird's 'head' (NGC 5394) and 'body' (NGC 5395) are a pair of galaxies suffering a near collision and stirring up gravitational trouble in each other.

The gravity of the larger galaxy has pulled out two long streamers from its companion, to create the heron's 'neck' and 'beak', and forced the stars in its centre to form an oval-shaped 'eye' – leading astronomers to classify NGC 5394 as an 'ocular galaxy' (a subclass of spiral galaxies).

Tom Williams's intimate portrait shows that the smaller galaxy has also ruffled the feathers in the heron's body, stirring up dark lanes of dusty gas where new stars are springing into life.

You'll find the Heron Galaxy in **Canes Venatici**, not far from a more famous galactic close encounter, the **Whirlpool Galaxy** (M51) and its companion NGC 5195, though the Heron is eight times further away – at 160 million light years – and 40 times fainter.

SEPTEMBER'S CALENDAR

SUNDAY	MONDAY	TUESDAY	WEDNESDAY	THURSDAY	FRIDAY	SATURDAY
			1	2	3	4 Moon near Mars
5	6 Moon near Antares	7 7.31 pm First Quarter Moon near Antares	8	9	10	11
12	13	14	15	16 0.03 am Full Moon	17	18 Moon near Saturn
19 Mare Orientale visible on Moon	20	21 Moon occults the Pleiades (am)	22	23 11.20 am Last Quarter Moon; Autumn Equinox	24 Mercury E elongation	25
26	27 Neptune closest to Earth	28 Moon near Jupiter (am); Neptune opposition	29	30 3.36 am New Moon		

SPECIAL EVENTS

- **4 September:** the crescent Moon lies to the lower left of Mars.
- **18 September:** Saturn is below the Moon.
- **19 September, 10 pm:** our best view of Mare Orientale – on the Moon's far side – since 2021 (see Topic and Chart 9a).
- **21 September, 6 am:** the Moon begins to occult the Pleiades just as the sky is brightening.
- **23 September, 7.02 am:** nights become longer than days as the Sun moves south of the Equator at the Autumn Equinox.
- **24 September:** Mercury reaches its greatest evening separation from the Sun, but is not visible in the bright dusk twilight.

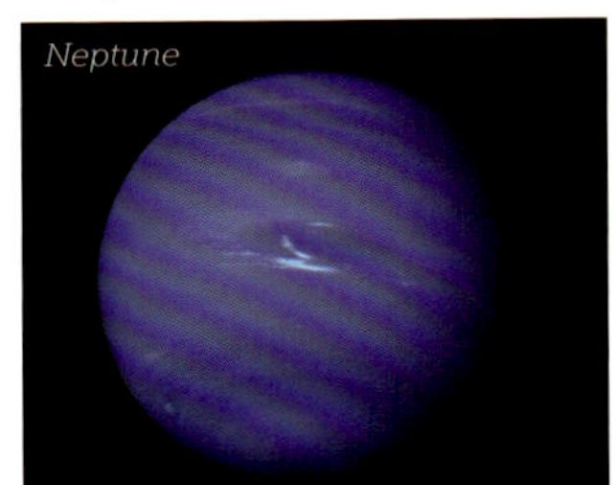
Neptune

- **27 September:** Neptune is closest to the Earth at 4,319 million km (see Planet Watch).
- **28 September, before dawn:** a lovely sight in the morning sky, as a narrow crescent Moon teams up with Jupiter (Chart 9b).
- **28 September:** Neptune is opposite the Sun in the sky (see Planet Watch).

SEPTEMBER'S PLANET WATCH

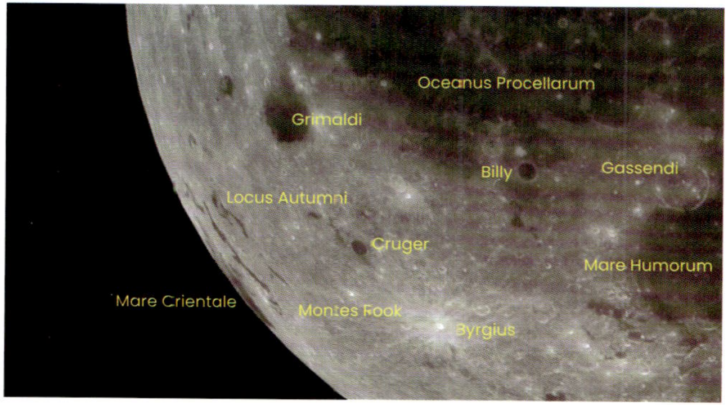

9a 19 September, 10 pm. The Moon's Mare Orientale comes into view.

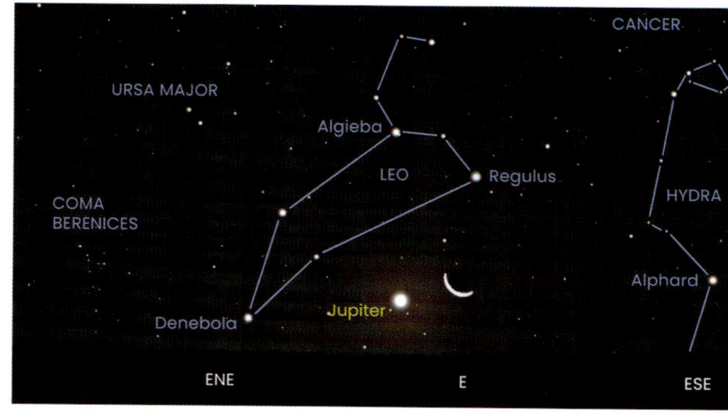

9b 28 September, 5.30 am. The crescent Moon near Jupiter.

- Very low in the south-west after sunset, you may catch a glimpse of **Mars**, now a mere shadow of its magnificence earlier this year as it shines at a measly magnitude +1.4. The Red Planet is setting around 8.30 pm, and travels from Virgo to Libra during the month, with the Moon nearby on 4 September.
- **Saturn** is clearing the eastern horizon about 8.30 pm, and shines at magnitude +0.2 in Pisces. The Moon lies nearby on 18 September.
- To its right, **Neptune** too lies in Pisces. At its closest to Earth this month (27 September) and opposite to the Sun (28 September), the outermost planet is visible all night and it at its brightest this year. Unfortunately, that amounts to a magnitude of +7.7, meaning that it's too faint to be seen with the unaided eye. Binoculars will show you this distant world looking like a faint star; you'll need telescopic power to see even a hint of the planet's bluish globe.
- **Uranus**, in Taurus – just above Aldebaran and the Hyades – is rising around 9.30 pm and shines at magnitude +5.7.
- Giant planet **Jupiter** reappears in the morning sky in the second half of September. It rises about 5.30 am in Leo. A thin crescent Moon lies to the right of Jupiter on the morning of 28 September (Chart 9b).
- You may catch **Mercury** very low in the west after sunset on the first few evenings of September, but it rapidly sinks into the dusk and is not visible when it reaches its greatest separation from the Sun on 24 September.
- **Venus** is lost in the Sun's glare this month.

- The sky at 11 pm in mid-October, with Moon positions at three-day intervals either side of Full Moon.
- The star positions are also correct for midnight at the beginning of October, and 9 pm at the end of the month (after the end of BST and IST).
- The planets move slightly relative to the stars during the month.

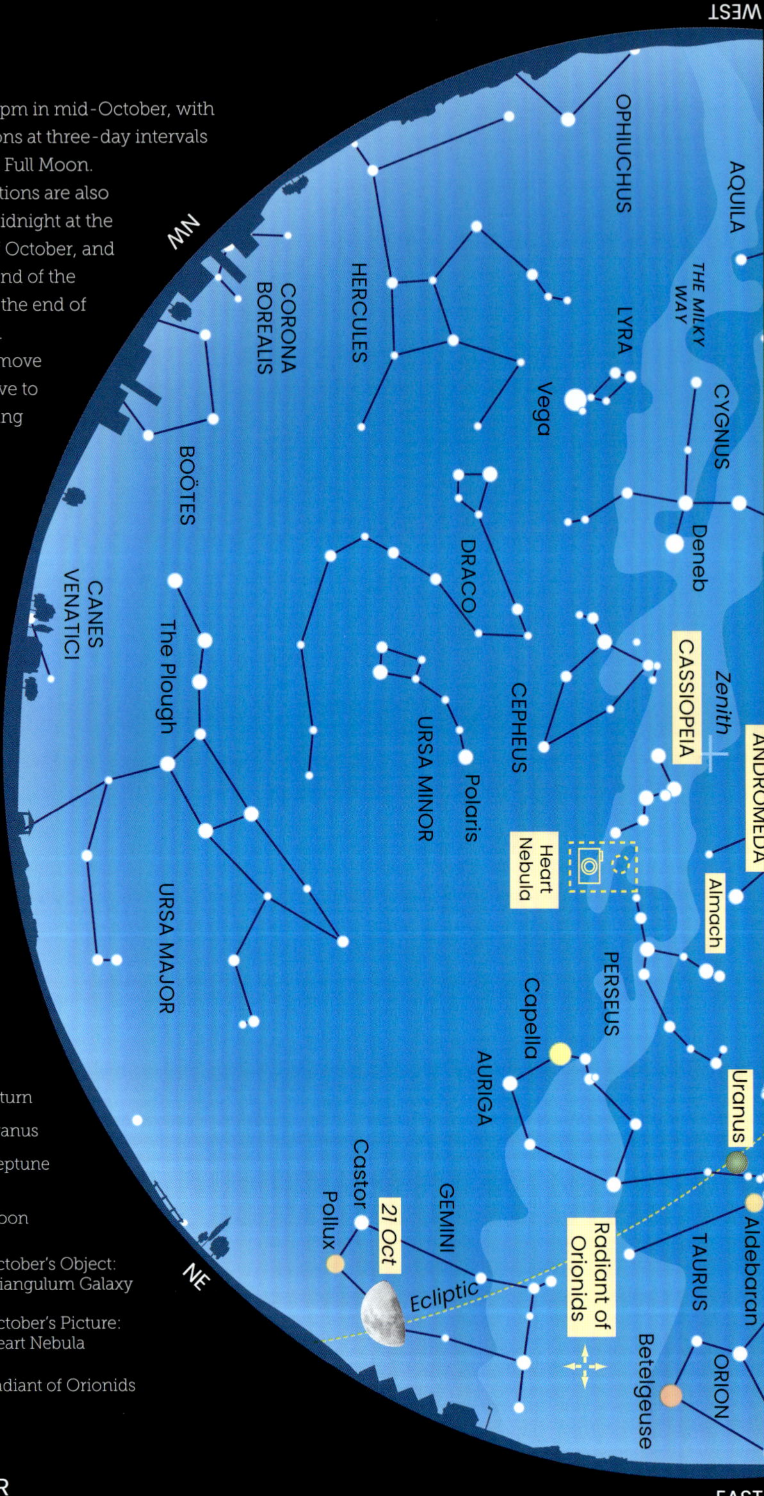

- Saturn
- Uranus
- Neptune
- Moon
- October's Object: Triangulum Galaxy
- October's Picture: Heart Nebula
- Radiant of Orionids

OCTOBER

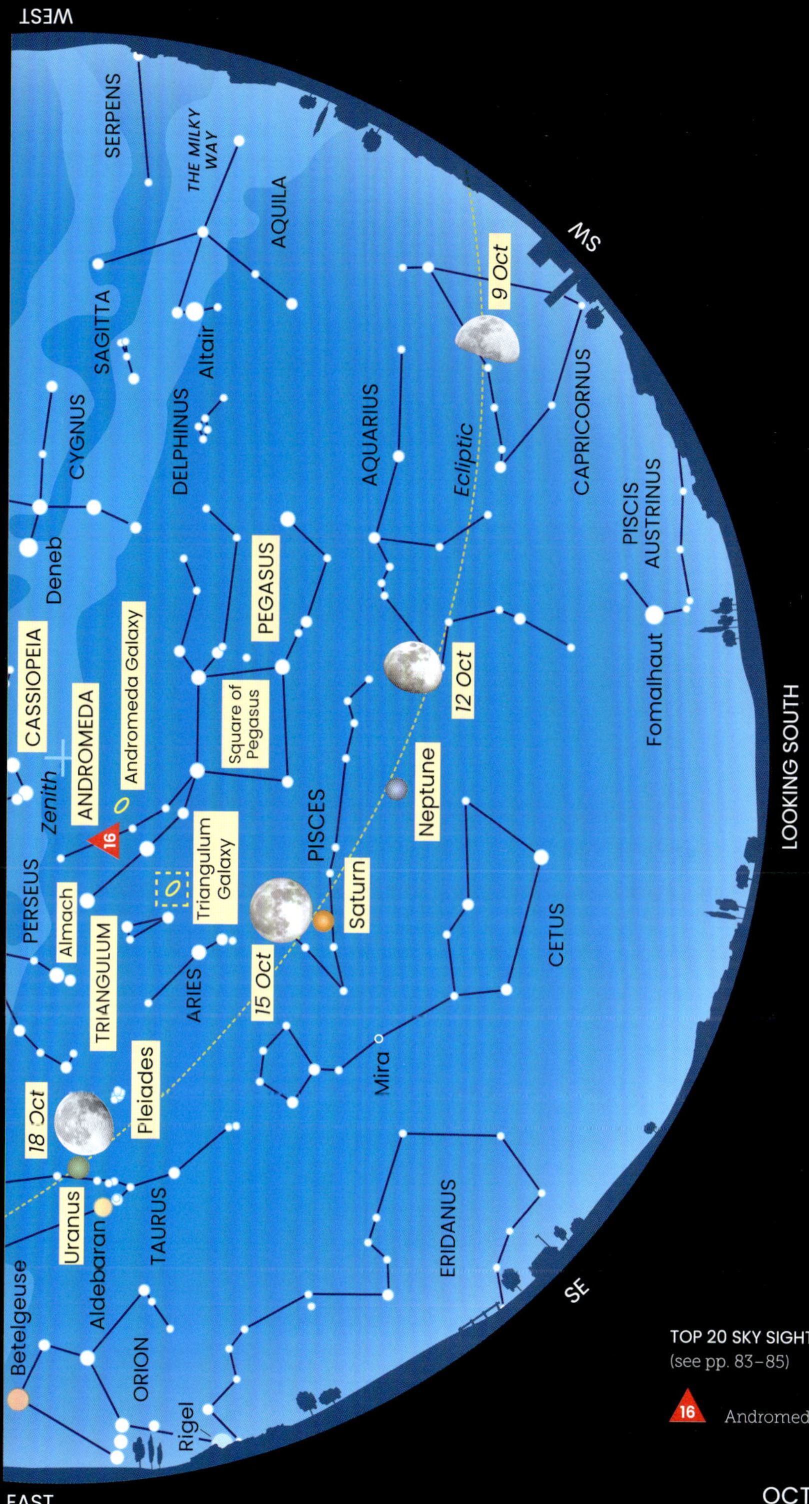

TOP 20 SKY SIGHTS
(see pp. 83–85)

16 Andromeda Galaxy

Saturn is king of the evening sky, at its brightest and closest this month. The constellations surrounding it are pretty drab: neither the **Square of Pegasus** nor **Andromeda** are guaranteed to thrill. But look to the east, where the brilliant lights of winter are starting to appear, spearheaded by the beautiful star cluster of the **Pleiades**.

OCTOBER'S CONSTELLATION

It may look like a simple line of stars linked to **Pegasus**, but the ancient Greeks saw **Andromeda** as a young princess chained to a rock, about to be devoured by a sea monster as a punishment for her mother, **Cassiopeia**, boasting her daughter was more beautiful than any of the sea nymphs.

Andromeda contains some surprising delights. **Almach**, the star at the left-hand end of the line, is a lovely sight in small telescopes: this beautiful double star comprises a yellow supergiant shining 2000 times brighter than the Sun, and a fainter bluish companion which is in fact triple.

But the glory of Andromeda is its great galaxy, beautifully placed on October nights. Lying above the line of stars, the **Andromeda Galaxy** is the most distant object easily visible to the unaided eye, a mind-boggling 2.5 million light years away: it's best seen in binoculars or a low-power telescope. Containing 400 billion stars, the Andromeda Galaxy is the biggest member of the Local Group of galaxies.

OBSERVING TIP

The Andromeda Galaxy is often described as the furthest object 'easily visible to the unaided eye'. It can be elusive, though – especially if you are suffering from light pollution. The trick is to memorise Andromeda's pattern of stars, and then to look slightly to the side of where you expect the galaxy to be. This technique – called 'averted vision' – causes the image to fall on the outer region of your retina, which is more sensitive to light than the central region that's evolved to discern fine details. Averted vision is also crucial when you want to observe the faintest nebulae or galaxies through a telescope.

OCTOBER'S OBJECT

The **Triangulum Galaxy** (M33) is the third largest member of the Local Group (after the Andromeda Galaxy and the Milky Way), and you can find it not far from Andromeda in the distinctive little star pattern of **Triangulum** (the Triangle). On the darkest of nights, you can just spot M33 with the naked eye, but I'd recommend using binoculars to find it.

M33 is home to 40 billion stars and – unlike Andromeda – it's a hotbed of star formation. Seeing any detail in this scruffy little spiral is a challenge, though, because its surface brightness is very low and – at 2.8 million light years – it lies slightly further away than Andromeda. Even with a large telescope, it's better to use a low magnification for the best views.

Take long-exposure photographs, and you'll uncover a multitude of ruby red nebulae studding the galaxy's arms. The biggest of these nebulae, NGC 604, is 60 times wider than the Orion Nebula and contains over 200 massive stars.

OCTOBER'S TOPIC: FRED HOYLE (1915–2001)

'We are stardust' is now such a common phrase that it even crops up in pop songs. The scientist who discovered this remarkable fact, Fred Hoyle, was a blunt Yorkshireman who was a radar pioneer during World War II, and then became fascinated by the origin of the elements.

Astronomers already knew that a star like the Sun shines because nuclear reactions convert hydrogen to helium in its core. In the 1950s, Hoyle assembled a group of scientists – known as B^2FH after their initials (Geoffrey and Margaret Burbidge, Willy Fowler and Hoyle) – who calculated how the intense inferno inside the most massive stars and the violence of a supernova explosion can produce all the chemical elements in the Universe.

Hoyle also promoted a controversial alternative to the Big Bang (a term that Hoyle coined): in his now discarded Steady State theory, the Universe has always existed and atoms of hydrogen are created as galaxies speed apart.

OCTOBER'S PICTURE

It was a cloudy night for astrophotographer Tim White; but he was ecstatic rather than downhearted – because these clouds weren't in Earth's atmosphere, but 7500 light years away in the centre of the **Heart Nebula**. As its name suggests, this nebula – located in Cassiopeia – is roughly shaped like a human heart, and Tim's close-up captures just the division between the two sides of the cosmic organ.

The colours show light emitted by hydrogen, sulphur and oxygen gas – but the view is dominated by the dark jagged silhouettes of denser gas clouds laced with dust particles that block out any light from passing through. Tim has removed the stars from this image to accentuate the wild shapes of these stygian cosmic clouds, totally different from the bright, watery clouds in the Earth's atmosphere.

While terrestrial clouds are shaped by winds and rising currents of warm air, producing streaks and fluffy shapes, the clouds in Cassiopeia are moulded by radiation from hot young stars. This intense energy blasts away the cloud's surface, giving rise to jagged shapes reminiscent of iron being eaten away by corrosion.

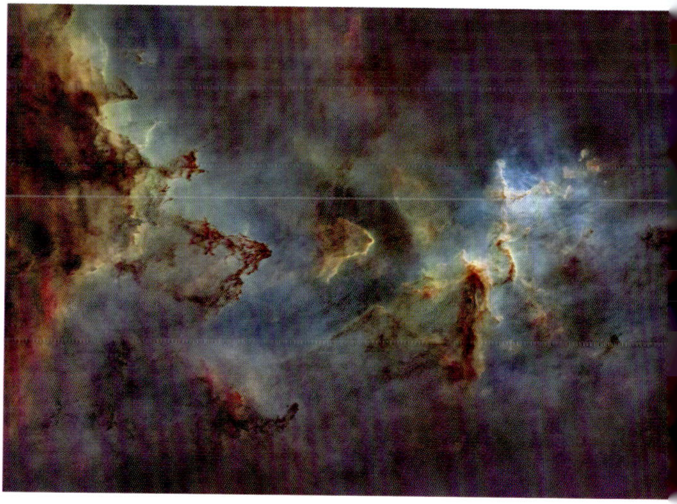

From his home in Devon, Tim White captured the heart of the Heart Nebula with a 235-mm Celestron EdgeHD Schmidt–Cassegrain telescope equipped with a ZWO ASI2600MM mono camera through Antlia 2.5-nm filters passing light from hydrogen (H alpha), oxygen ([OIII]) and sulphur (SII) for a total exposure of 18 hours.

OCTOBER'S CALENDAR

SUNDAY	MONDAY	TUESDAY	WEDNESDAY	THURSDAY	FRIDAY	SATURDAY
31 BST and IST end					1	2
3 Moon near Mars	4 Moon near Antares	5	6	7 12.47 pm First Quarter Moon	8	9
10	11	12	13	14	15 2.47 pm Full Moon near Saturn	16
17	18 Saturn opposition	19	20	21 Orionids	22 Orionids (am); 6.29 pm Last Quarter Moon	23
24	25	26 Moon near Jupiter (am)	27	28	29 2.36 pm New Moon	30

SPECIAL EVENTS

- **3 October:** a narrow crescent Moon joins Mars very low on the south-western horizon as the sky grows dark.
- **4 October:** the crescent Moon hangs right underneath Antares.
- **15 October:** the Full Moon lies directly above Saturn (Chart 10a); with the ringworld nearly at opposition, this means that four astronomical bodies – the Sun, the Earth, the Moon and Saturn – are almost exactly in a straight line.
- **18 October:** Saturn lies opposite the Sun in the sky, and closest to Earth at 1246 million km.
- **Night of 21/22 October:** high-speed debris from Halley's Comet smashes into the Earth's atmosphere as the annual Orionid meteor shower, but this year all but the brightest shooting stars are erased by the light of the gibbous Moon lying near its radiant.
- **26 October, before dawn:** a lovely early morning sight as the crescent Moon lies to the right of Jupiter (Chart 10b).
- **31 October, 2 am:** the end of British Summer Time and Irish Standard Time.

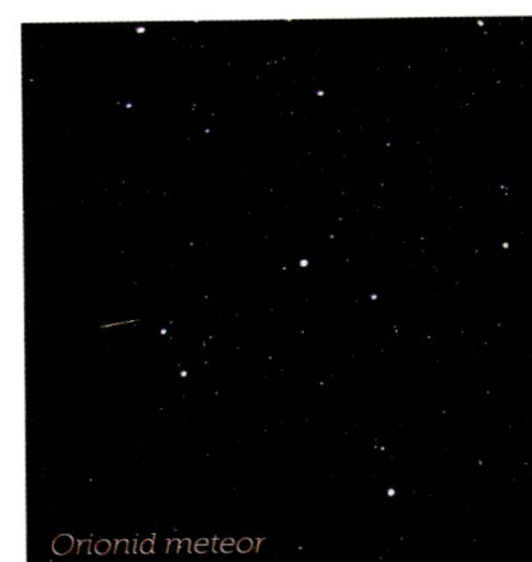
Orionid meteor

OCTOBER'S PLANET WATCH

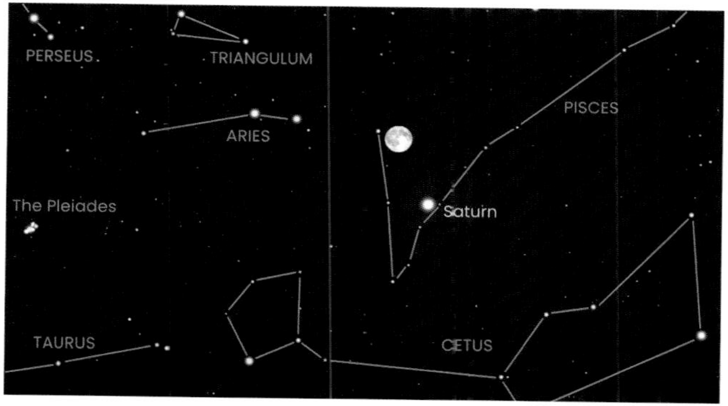

10a *15 October, 9 pm. The Full Moon passes above Saturn, as the Sun, Earth, Moon and Saturn line up.*

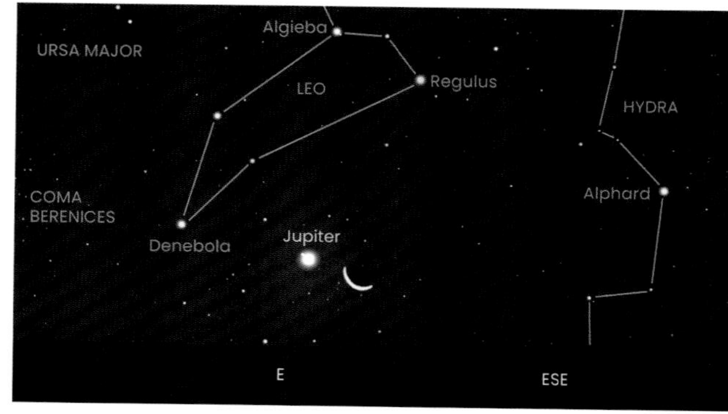

10b *26 October, 4.30 am. The crescent Moon near Jupiter.*

• **Mars** continues to hug the south-western horizon this month, moving from Libra through Scorpius to Ophiuchus. The Red Planet sets around 7.30 pm, and shines at magnitude +1.4.

• In the last week of October, **Venus** appears on the horizon to the lower right of Mars: blazing at magnitude −3.9, it sets about 6.30 pm.

• Lying in Pisces, **Saturn** is the leading light of the evening sky, visible all night long and resplendent in a region of sky that contains only faint stars. The planet is at its brightest this year (magnitude +0.1) as it reaches opposition and its closest point to Earth on 18 October. Grab a telescope to view its famous rings, now appearing more open as our perspective changes from an edge-on position in 2025 to a steeper viewing angle. Also look out for the planet's largest moon, Titan, as it orbits Saturn every 16 days. The Full Moon passes Saturn on 15 October (Chart 10a).

• **Neptune** (magnitude +7.7) also lies in Pisces and sets about 6 am.

• In Taurus, **Uranus** rises around 8 pm and shines at magnitude +5.6.

• **Jupiter**, in Leo, rises about 4 am at a brilliant magnitude −1.8. The Moon is nearby on the morning of 26 October (Chart 10b).

• On the last few mornings of October, you may catch **Mercury** (magnitude 0.0) down on the horizon to the lower left of Jupiter, and rising around 6 am.

- The sky at 10 pm in mid-November, with Moon positions at three-day intervals either side of Full Moon.
- The star positions are also correct for 11 pm at the beginning of November, and 9 pm at the end of the month.
- The planets move slightly relative to the stars during the month.

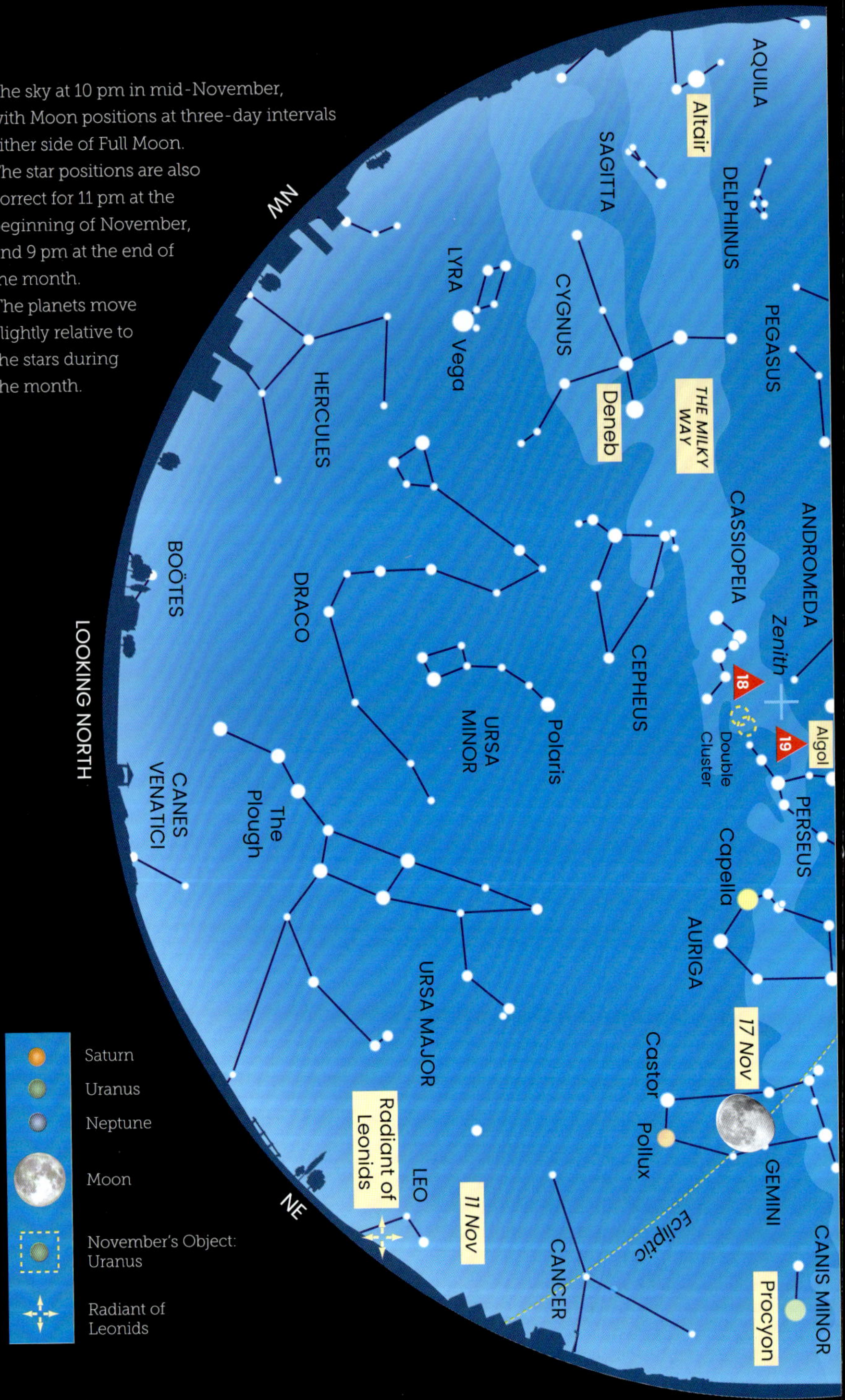

NOVEMBER

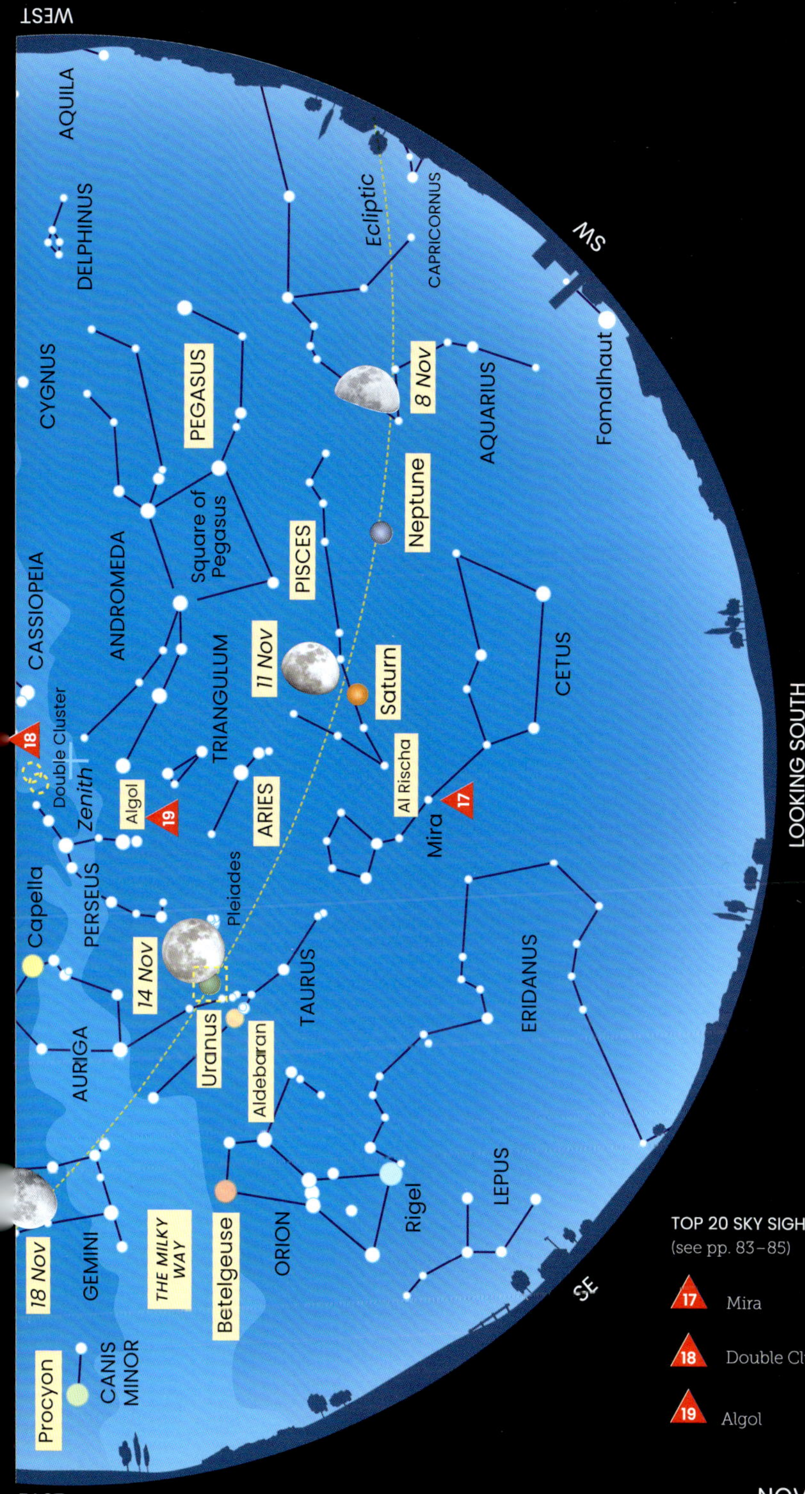

TOP 20 SKY SIGHTS
(see pp. 83–85)

- 17 Mira
- 18 Double Cluster
- 19 Algol

Look out for a planetary pas-de-deux between Venus and Mars, with the crescent Moon joining them for a couple of lovely photo ops. Also, check out the **Milky Way**: sweep along the glowing band with binoculars, and you'll find it studded with star clusters and nebulae.

NOVEMBER'S CONSTELLATION

Pisces is typical of the autumn constellations – faint and straggly, as it loops around the sky below **Pegasus**. Many ancient civilizations – from the Babylonians to the Egyptians and the early Christians – discerned a pair of fishes in this star pattern. To the ancient Greeks, they were the goddess Aphrodite and her son Eros, converted into piscine form and tied by a cord on their scaly tails, in order to escape from the monster Typhon.

The central star **Al Rischa** (whose name means 'cord' in Arabic) – lying to the lower left of Saturn this month – was discovered to be double, by William Herschel in 1779. The pair circle each other every 720 years, and currently appear very close together.

Pisces is the location of the Vernal Equinox – the point where the Sun crosses the celestial equator on its way from the southern to the northern hemisphere.

OBSERVING TIP

With Christmas on the way, you may be thinking of buying a telescope as a present for a budding stargazer. Beware! Unscrupulous websites and mail-order catalogues often advertise small telescopes that boast huge magnifications, but all they do is show you a large blurry image. To see planets and stars sharply, you need to use a magnification no more than twice the diameter of the lens or mirror in millimetres. So if you see an advertisement for a 75-mm telescope, beware of any claims for a magnification greater than 150 times.

NOVEMBER'S OBJECT

If you're sharp-sighted and have extremely dark skies, you stand a chance of spotting **Uranus** – the most distant planet visible to the unaided eye – at its closest to Earth on 30 November. Discovered in 1781 by William Herschel, Uranus circles the Sun every 84 years. It has 13 faint narrow rings and at least 29 moons.

Four times the diameter of the Earth, Uranus orbits the Sun on its side (probably the result of a collision in its infancy). When the Voyager 2 spacecraft zipped past Uranus in 1986 – beaming back our only close-up views – it revealed a bland, featureless world, but later telescope observations have shown vast stormy clouds changing with the seasons.

Uranus consists largely of a vast watery ocean surrounding a hot rocky core. The latest theories posit a hailstorm of diamonds dropping down through the water, to land in a sea of liquid diamond that wraps around the planet's core.

NOVEMBER'S TOPIC: STAR NAMES

Why do the brightest stars have such strange names such as **Aldebaran**, Antares or **Altair**? While many of the star patterns were drawn up by the Babylonians or Minoans before 1000 BC, most star names trace back to Greek astronomers around AD 150 (see March's Topic).

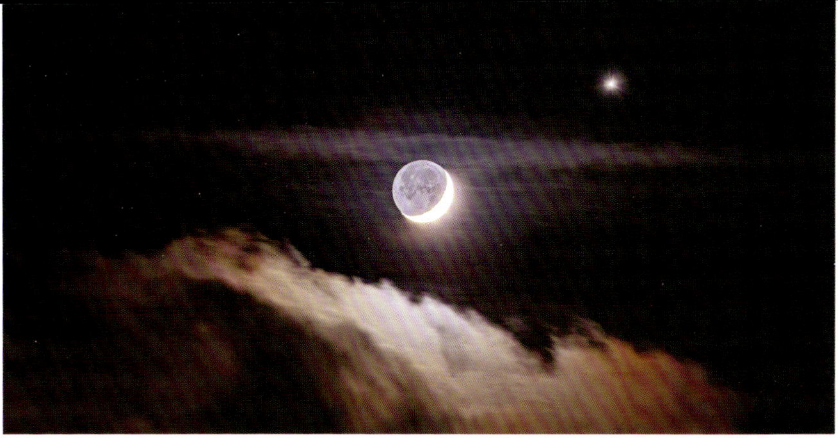

'Knowing the conjunction was happening,' explains Jonathen Harty, 'I waited for a gap in the clouds' – with his Fujifilm, X-T5 mirrorless camera plus Fujifilm 50-mm–140-mm zoom lens at full zoom, mounted on a tripod. And when that gap came, he took ten shots at various shutter speeds, to try to expose both the crescent Moon and the Da Vinci glow. He succeeded with a 2-second exposure at f/2.8 and ISO 800.

When it comes to Antares (see June's Constellation) the name is pure Greek, meaning 'rival of Ares [Mars]' because its colour vies with that of the Red Planet. Sirius comes from the Greek for 'scorcher' as they believed this brilliant star's rays added to the Sun's heat in summer.

But the Arabs were mainly responsible for the star names we use today. Between the sixth and tenth centuries AD, they translated many of the earlier names into their own language – hence the ubiquity of the 'al' (Arabic for 'the') as the first syllable. **Algol** means 'the demon' – possibly because they noticed its brightness seems to 'wink' every few days, while Aldebaran and Altair are the 'the follower' and 'the flying eagle'. **Deneb** also has Arabic roots, meaning 'the tail (of the flying bird)'.

The most famous star in the sky has to be **Betelgeuse**, but its name was long a mystery. The 'B' in Betelgeuse turns out to be a mistransliteration of an Arabic letter, and the name just means 'Orion's hand'.

NOVEMBER'S PICTURE

Nothing is more magical than the crescent Moon pairing up with Venus, low on the horizon after sunset. This month we're treated to two Moon–Venus conjunctions – on 1 and 30 November – with Mars thrown in as a bonus.

In Jonathen Harty's evocative shot, the Moon's bright crescent encloses a ghostly view of the remainder of the lunar surface, thanks to 'earthshine' – sunlight that's been reflected off the Earth and back onto the Moon's unlit surface. Traditionally known as 'the Old Moon in the New Moon's arms', earthshine was first explained correctly by Leonardo da Vinci (accomplished scientist as well as artist and inventor) in 1510, and it's sometimes called the Da Vinci glow.

Venus appears so bright in our night sky – outshone only by the Moon – because it's covered in brilliant white clouds that reflect 75 per cent of the sunlight shining on the planet. (In contrast, the Moon's surface reflects only 12 per cent of the Sun's light, and the Earth 30 per cent.)

The night Jonathen took this photo, clouds in the Earth's own atmosphere scudding across the foreground added an extra layer of depth and artistry.

NOVEMBER'S CALENDAR

SUNDAY	MONDAY	TUESDAY	WEDNESDAY	THURSDAY	FRIDAY	SATURDAY
	1 Moon near Mars and Venus	2	3	4 Mercury W elongation	5	6 8.00 am First Quarter Moon
7	8	9	10	11 Moon near Saturn	12	13
14 3.26 am Full Moon occults the Pleiades	15	16	17 Leonids	18 Leonids (am)	19	20 Moon near Regulus
21 0.48 am Last Quarter Moon	22 Moon between Jupiter and Regulus (am)	23 Moon near Jupiter (am)	24 Venus very near Mars	25 Moon near Spica (am); Venus very near Mars	26	27
28 3.24 am New Moon	29	30 Moon near Venus and Mars; Uranus opposition				

SPECIAL EVENTS

- **1 November:** very low in the south-west, the crescent Moon lies to the left of Venus in the evening twilight; fainter Mars is in between.
- **4 November:** Mercury reaches its greatest morning separation from the Sun (see Planet Watch).
- **11 November:** the bright 'star' below the Moon is the planet Saturn.
- **14 November, 5–7.30 pm:** the Full Moon moves across the lower part of the Pleiades, occulting some of its stars (Chart 11a). The brilliance of the lunar orb means you'll need binoculars or a telescope to spot the stars in the cluster.
- **Night of 17/18 November:** the maximum of the Leonid meteor shower – fiery fragments of Comet Tempel–Tuttle – is spoilt by bright moonlight.
- **22 November, early hours:** you'll find Jupiter to the lower left of the Moon, with Regulus above.
- **23 November, early hours:** Jupiter lies directly over the crescent Moon.
- **24 November:** Mars lies just 20 arcminutes above Venus.
- **25 November:** only 27 arcminutes separate Venus and Mars (to its right).
- **30 November:** low in the south-west after sunset, the crescent Moon is right next to Venus, with Mars to the right (Chart 11b).
- **30 November:** Uranus is opposite to the Sun, and closest to the Earth at 2,749 million km (see Planet Watch).

NOVEMBER'S PLANET WATCH

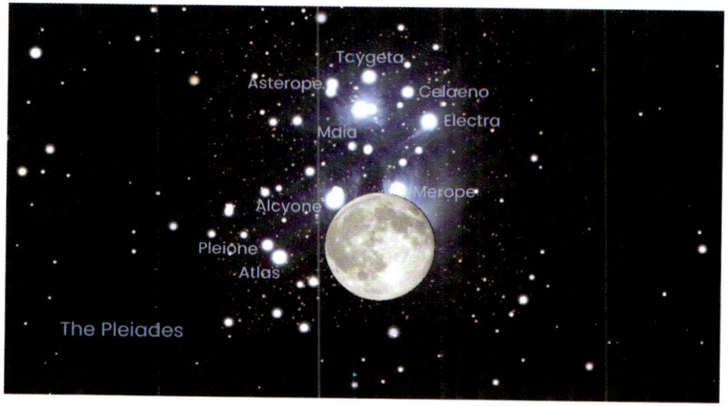

11a *14 November, 6 pm. The Moon occults the Pleiades.*

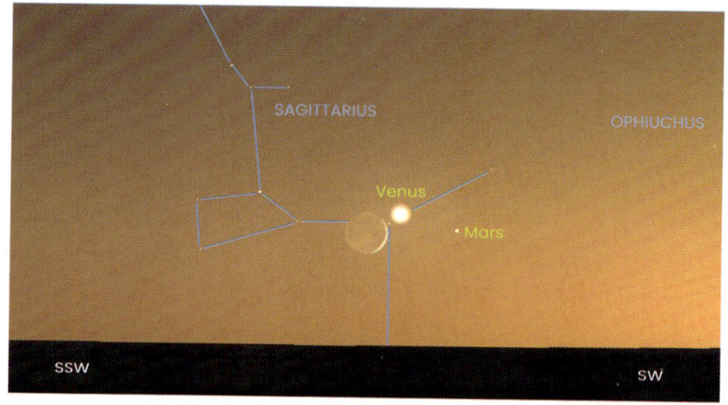

11b *30 November, 4.30 pm. The Moon next to Venus and Mars.*

- **Venus** is low in the dusk twilight, to the south-west, setting around 5.30 pm and blazing at magnitude −3.9. It lies near to **Mars**, which is a hundred times fainter at magnitude −1.3 and is a challenge to spot against the twilight glow (binoculars are a great help here): the Red Planet tracks from Ophiuchus to Sagittarius during November.
- On 1 November, the crescent Moon lies to the left of Venus, with Mars in between. Venus is heading towards the Red Planet, and it passes Mars more closely than the diameter of the Moon on 24 and 25 November (see Special Events). And the three nearest worlds – the Moon, Venus and Mars – assemble again in a spectacularly tight group on 30 November (Chart 11b).
- **Saturn** lies in Pisces. Shining at magnitude +0.2, the planet sets around 4.30 am. The Moon passes above it on 11 November.
- Also in Pisces, **Neptune** is a dim magnitude +7.7 and sets about 3 am.
- Visible all night long in Taurus, **Uranus** is at opposition and closest to the Earth on 30 November (see Special Events) when the planet is at its brightest this year – though, at magnitude +5.6, it's only just visible to the unaided eye. You can easily see Uranus in binoculars, and a moderate telescope will show the planet as a bluish-green disc.
- **Jupiter** (magnitude −1.9) rises around 1.30 am, crossing from Leo to Virgo. The Moon is nearby on the mornings of 22 and 23 November.
- **Mercury** is putting on its best morning display of the year during the first half of the month, reaching its greatest elongation from the Sun on 4 November. You'll find the innermost planet low in the east before dawn, at magnitude −0.8, and rising about 5.20 am.

- The sky at 10 pm in mid-December, with Moon positions at three-day intervals either side of Full Moon.
- The star positions are also correct for 11 pm at the beginning of December, and 9 pm at the end of the month.
- The planets move slightly relative to the stars during the month.

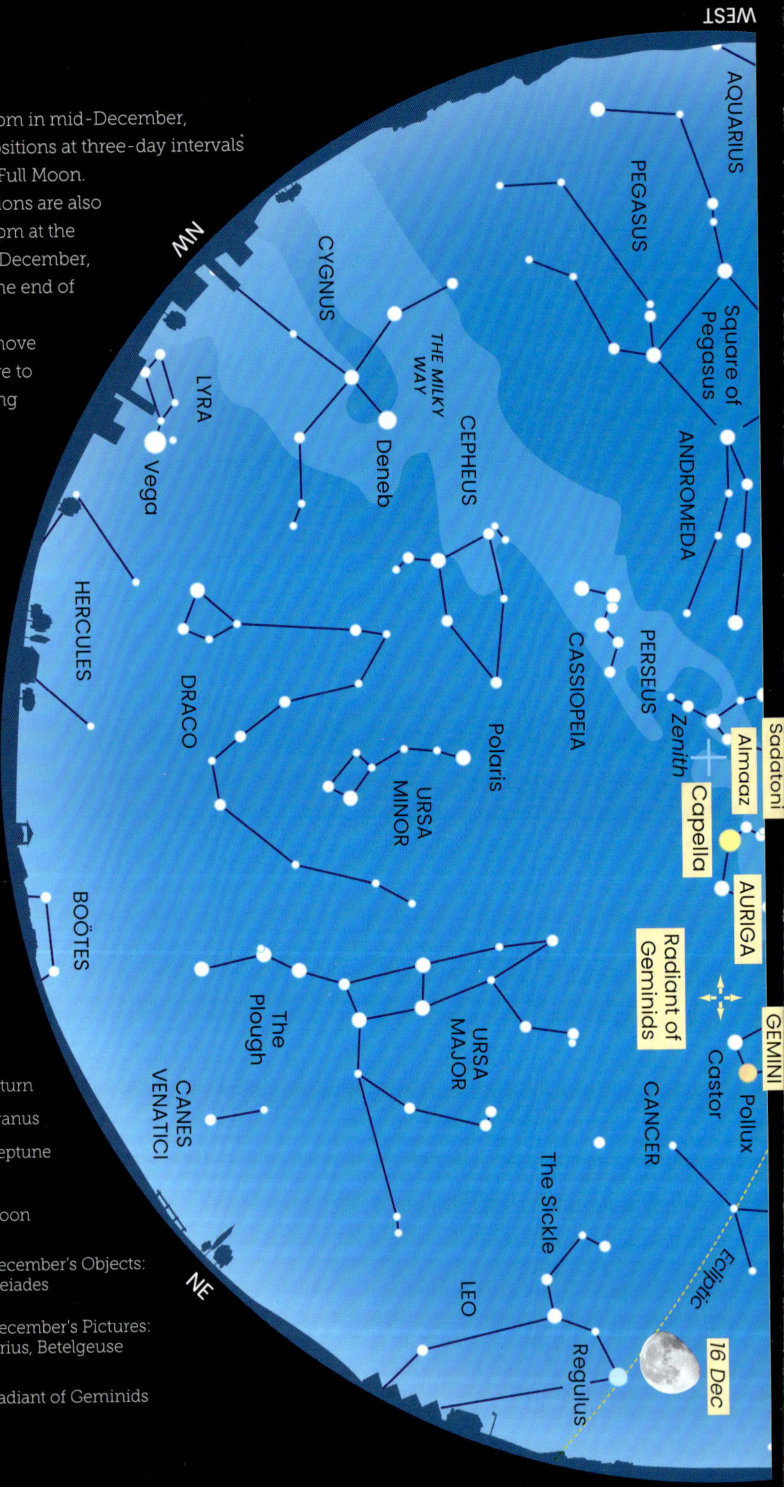

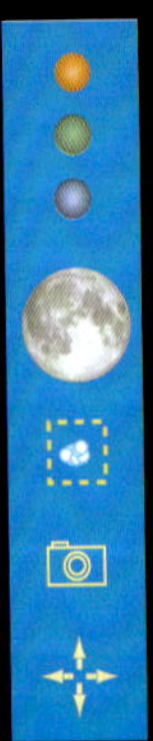

DECEMBER

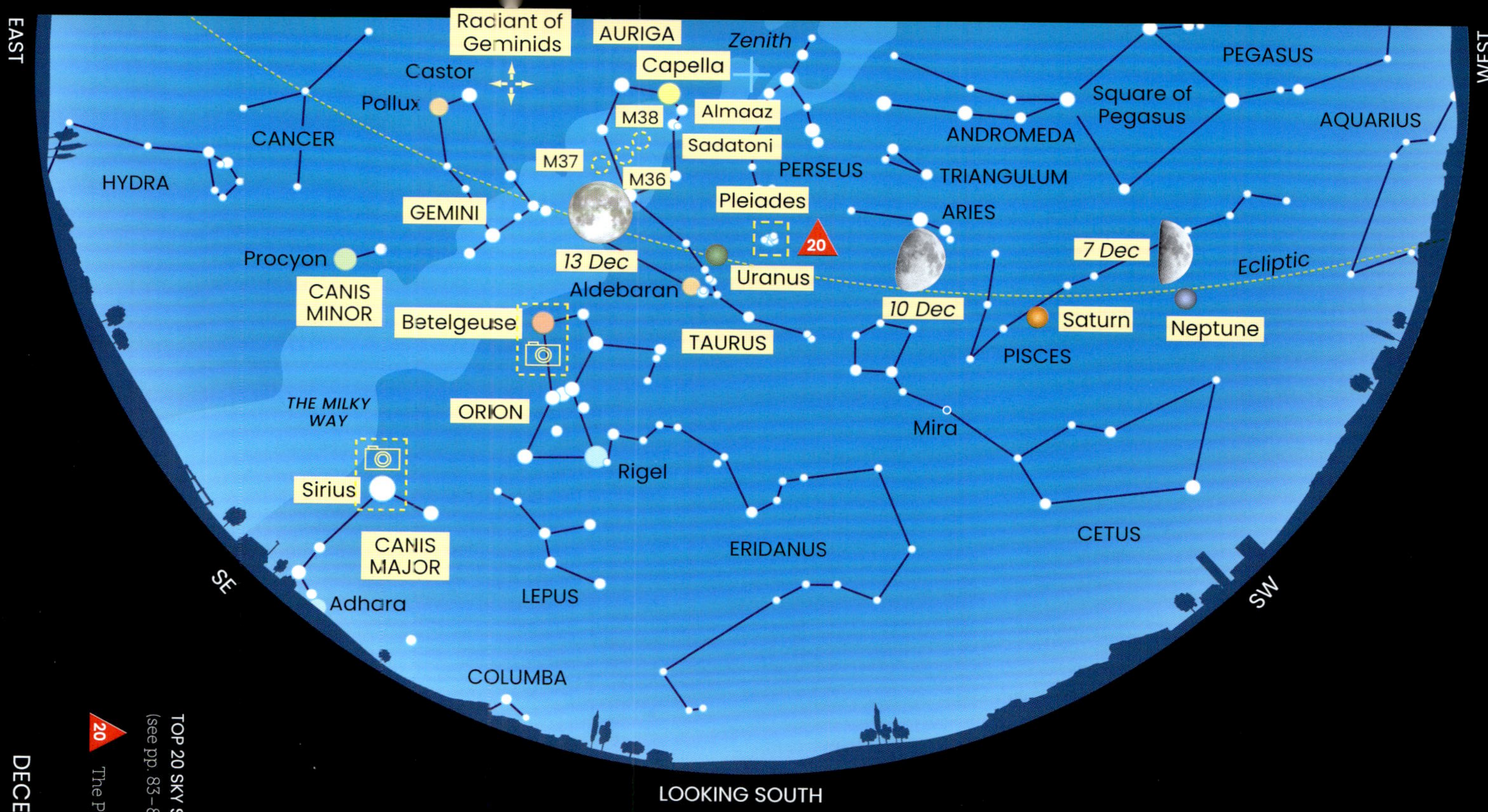

TOP 20 SKY SIGHTS
(see pp. 83–85)

20 The Pleiades

We have a 'Christmas Star', as Venus appears magnificently in the evening sky. The stars, too, are putting on a festive show, featuring **Orion** with his hunting dogs **Canis Major** and **Canis Minor**, co-starring **Taurus** (the Bull), the hero twins of **Gemini** and the charioteer **Auriga**.

DECEMBER'S CONSTELLATION

Sparkling overhead, **Auriga** (the Charioteer) is named after the Greek hero Erichthonius, who invented the four-horse chariot. The ancient Babylonians, on the other hand, saw Auriga as a shepherd's crook.

Capella, the sixth-brightest star in the sky, means 'the little nanny goat'. It actually consists of two stars, each 75 times brighter than the Sun, orbiting more closely than the Earth circles the Sun.

Nearby, you'll find two eclipsing binaries: stars that regularly change in brightness because a companion passes in front. **Sadatoni** is an orange star eclipsed every 2 years 8 months by a blue partner. **Almaaz** has a companion star that's completely hidden by a huge dark disc of cosmic dust swirling around it. Every 27 years, this disc moves in front of Almaaz and eclipses it for two years.

And bring out your binoculars (better still, a small telescope) to view three very pretty star clusters within the outline of Auriga, **M36**, **M37** and **M38**.

DECEMBER'S OBJECT

One of the delights of autumn and winter is a celestial gem in the constellation Taurus, the gently glowing **Pleiades** star cluster. My favourite description comes not from an astronomer, but from the pen of the Victorian poet Alfred, Lord Tennyson: 'Many a night I saw the Pleiads, rising through the mellow shade,/Glitter like a swarm of fireflies tangled in a silver braid.'

Though the cluster is well known as the Seven Sisters, skywatchers typically see any number of stars but seven! Most people can pick out the six brightest stars, while very keen-sighted observers can discern up to 14 members.

These are just the most luminous in a group of at least 1000 stars, lying 444 light years away. The brightest stars in the Pleiades are hot and blue, and all the stars are young – around 100 million years old (about 2 per cent of the Sun's age).

DECEMBER'S TOPIC: THE CHRISTMAS STAR

Many people will be looking at brilliant Venus this month, and wondering what was the celestial wonder that fascinated the Magi from the East, and led them to Bethlehem?

We know that Jesus wasn't born in AD 1, but a few years earlier (the confusion in dates was due to a counting

OBSERVING TIP

Venus is glorious this month. Through a small telescope, you can make out its cloud-wreathed globe, three-quarters lit by the Sun. But don't wait for the sky to get totally dark. Seen against a black sky, Venus is so brilliant it's difficult to discern any details. You're best off viewing Venus soon after the Sun has set, when the Evening Star first becomes visible in the twilight glow. Through a telescope, the planet then appears less dazzling against a pale blue sky.

error by a later Christian monk). During this period, Chinese astronomers were assiduously recording the sky, and they reported nothing as dazzling as the star that appears on our Christmas cards.

Perhaps the portent was more subtle. The Magi were astrologers, and the 'star' might have been an unusual planetary tango in 7 BC, between Jupiter (the king of the planets) and Saturn (representing justice) in Pisces (the constellation of the Jews). Or perhaps it was a comet, that the Chinese observed for 70 days in 5 BC?

But none of these completely match the gospel account – especially that the star 'went before them, till it came and stood over where the young child was'.

Perhaps it's significant that the star is only mentioned by St Matthew (St Luke has the shepherds). His gospel uniquely stresses that the life and teachings of Jesus fulfil the predictions of the Old Testament. Starting with the prophecies 'a star will come out of Jacob' (Book of Numbers) and 'arise, shine, for your light has come' (Isaiah), Matthew knew that his nativity account had to feature a brilliant shining star.

DECEMBER'S PICTURE

'Twinkle, twinkle, little star...' Stars, as the lullaby says, scintillate romantically; planets, on the other hand, don't. We can see the difference graphically in Mary McIntyre's time exposures she calls 'star squiggles'. During a 3.6-second exposure, Mary deliberately wobbled her camera tripod so that each object traces out a looping path.

Pure white **Sirius** (left), the brightest star in the sky, flickers erratically in brightness and colour like a temperamental diamond – behaviour immortalised in its star squiggle.

Betelgeuse's squiggle (centre) shows the red giant's colour and brightness also scintillating, but this star is hardly emitting any light at the shorter wavelengths, so there's little white or blue in its trace.

Twinkling is entirely due to the Earth's turbulent atmosphere, which twists and breaks up starlight passing through. But the turbulence is small scale; and if you're looking at a planet the twinkling across its disc tends to cancel out, so a planet shines with a constant light, as you can seen in Jupiter's squiggle (right).

Mary McIntyre captured these 'star squiggles' from Oxfordshire using a Canon EOS 1100D camera with a Sigma 300-mm zoom lens at 300-mm focal length. The camera was set to ISO 400 and f/5.6, and she took 3.6-second exposures while deliberately wobbling the tripod.

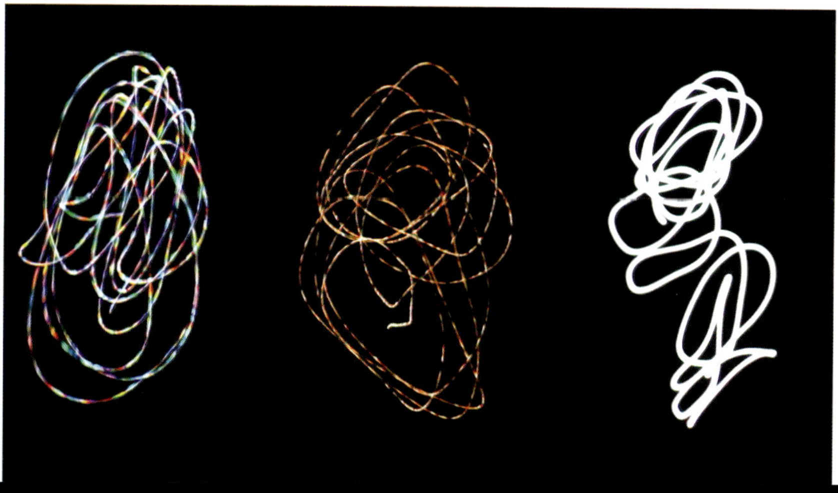

DECEMBER'S CALENDAR

SUNDAY	MONDAY	TUESDAY	WEDNESDAY	THURSDAY	FRIDAY	SATURDAY
			1	2	3	4
5	6 5.22 am First Quarter Moon	7	8 Moon near Saturn	9	10	11
12 Moon occults the Pleiades (am)	13 4.09 pm Full Moon; Geminids	14 Geminids (am); Geminids	15 Geminids (am)	16	17 Moon near Regulus	18
19	20 9.11 am Last Quarter Moon near Jupiter	21	22 Moon near Spica (am); Winter Solstice	23	24	25
26	27 8.12 pm New Moon	28	29 Moon near Mars	30 Moon near Venus	31	

SPECIAL EVENTS

- **8 December:** the Moon lies above Saturn.
- **12 December, 4–6.30 am:** the Moon moves over the lower parts of the Pleiades, hiding some of its brightest stars (Chart 12a).
- **Nights of 13/14 and 14/15 December:** the slow bright shooting stars of the Geminid meteor shower – dust grains from asteroid Phaethon – are all but washed out by the light of a Moon that's just past Full.

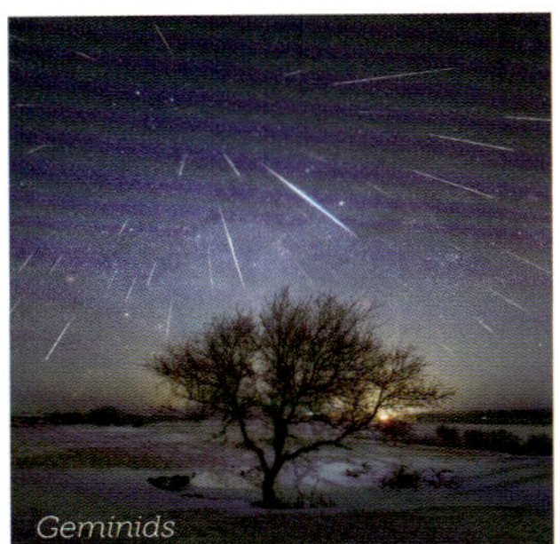

Geminids

- **20 December:** Jupiter lies to the left of the Last Quarter Moon.
- **22 December, 2.43 am:** the Winter Solstice, when the Sun reaches its lowest point in the sky and the northern hemisphere has the shortest day and the longest night.
- **29 December:** the narrowest crescent Moon hangs above Mars, deep in the evening twilight, with Venus well to the upper left.
- **30 December:** Venus and the crescent Moon form a stunning duo in the south-west after sunset (Chart 12b).

DECEMBER'S PLANET WATCH

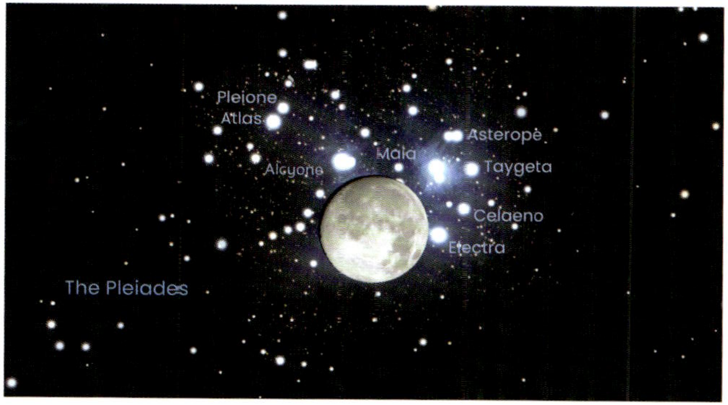

12a *12 December, 3.15 am. The Moon occults the Pleiades.*

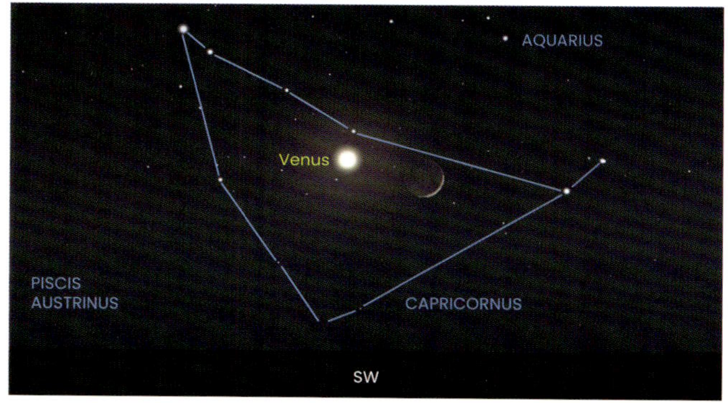

12b *30 December, 5.30 pm. The crescent Moon with Venus.*

- We have a glorious Evening Star this month, as **Venus** zooms rapidly upwards. By the end of the month the planet, at magnitude −4.0, is effulgent against a dark sky background, and sets as late as 7 pm. The crescent Moon makes a picturesque pair with Venus on 30 December (Chart 12b).

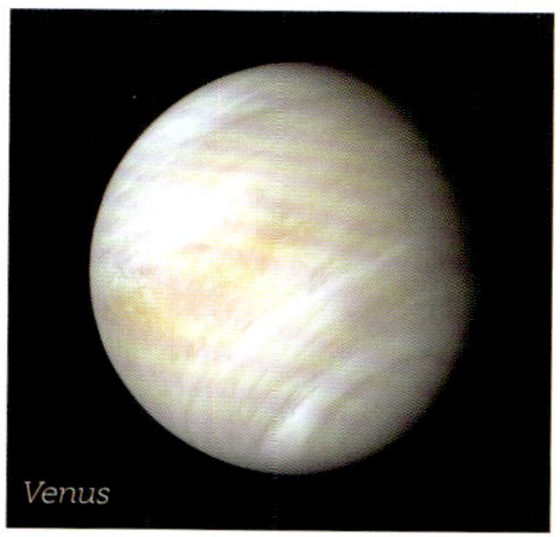

Venus

- **Mars** hangs low on the south-western horizon in the evening twilight all month, setting around 5.20 pm. The Red Planet shines at magnitude +1.2, moving through Sagittarius during December. The narrowest crescent Moon lies nearby on 29 December.
- Setting at 2.30 am, **Saturn** lies in Pisces at magnitude +0.4. The Moon is nearby on 8 December.
- Faint **Neptune** (magnitude +7.8), in Pisces, sets around 1 am.
- **Uranus** is setting about 6.30 am and shines at magnitude +5.6 in Taurus.
- Giant planet **Jupiter** lies in Virgo: shining at magnitude −2.2, it rises just before midnight. The Last Quarter Moon joins Jupiter on 20 December.
- **Mercury** is too close to the Sun to be seen this month.

Can you see the planets? It's a common question; and the answer is a resounding 'yes!' Some of our cosmic neighbours are the brightest objects in the night sky after the Moon. As they're so close, you can watch them getting up to their antics from night to night. And planetary debris – leftovers from the birth of the Solar System – can light up our skies as glowing comets and the celestial fireworks of a meteor shower.

THE SUN-HUGGERS

Mercury and Venus orbit the Sun more closely than our own planet, so they never seem to stray far from our local star: you can spot them in the west after sunset, or the east before dawn, but never all night long. At *elongation*, the planet is at its greatest separation from the Sun, though – as you can see in the diagram (right) – that's not when the planet is at its brightest. Through a telescope, Mercury and Venus (technically known as the *inferior planets*) show phases like the Moon – from a thin crescent to a full globe – as they orbit the Sun.

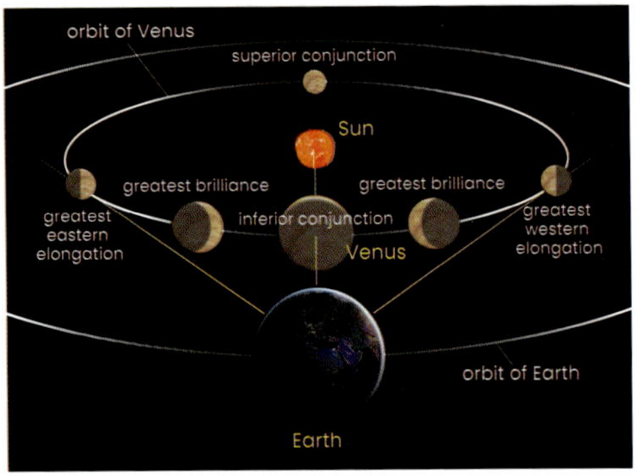

Venus (and Mercury) show phases like the Moon as they orbit the Sun.

Mercury

The innermost planet makes its first evening appearance in January–February and puts on a better performance in May; you may catch it again low in the dusk twilight in August–September. In terms of its morning appearances, Mercury is drowned out by bright twilight at its March elongation, but you can catch the innermost planet in July–August, and again in October–November when Mercury puts on its best show in the morning sky.

Maximum elongations of Mercury in 2027	
Date	Separation
3 February	18° east
17 March	28° west
28 May	23° east
15 July	21° west
24 September	26° east
4 November	19° west

Venus

Through to April we can enjoy Venus as the Morning Star. It's lost in the dawn twilight for several months, passing behind the Sun on 11 August and reappearing in the evening sky in October. The Evening Star passes very close to Mars on **24–25 November**.

Maximum elongation of Venus in 2027	
Date	Separation
3 January	47° west

WORLDS BEYOND

A planet orbiting the Sun beyond the Earth (known in the jargon as a *superior planet*) can be visible at all times of night, as we look outwards into the Solar System. It lies due south at midnight when the Sun, the Earth and the planet are all in line – a time known as *opposition* (see the diagram, right). Around this time the Earth lies nearest to the planet, although the date of closest approach (and the planet's maximum brightness) may differ by a few days because the planets' orbits are not circular.

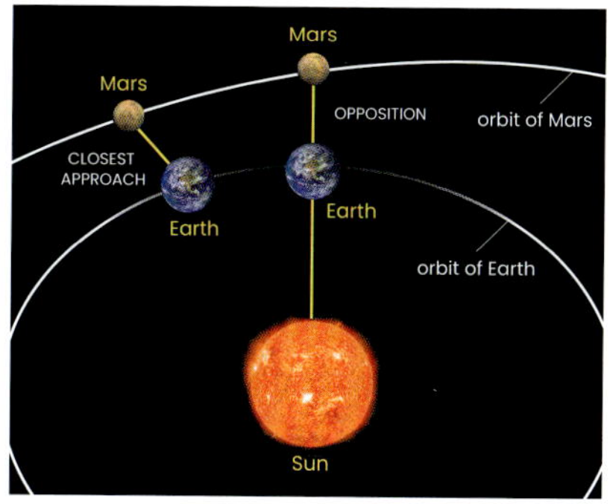

Mars (and the outer planets) line up with the Sun and Earth at opposition, but they are brightest at their point of closest approach.

Mars

The Red Planet is at its most prominent for two years, tangoing with Jupiter and Regulus for the first half of 2027. Mars is opposite the Sun on **19 February** and closest to us on **20 February**, when it passes above Regulus. Fading all the time, Mars moves back over Regulus on **14 May**, and then flies by Spica on **23–24 August** and Venus on **24–25 November**.

Where to find Mars	
January–June	Leo
July–mid-September	Virgo
Mid-September–mid-October	Libra
Late October	Scorpius
Early November	Ophiuchus
Mid-November–December	Sagittarius

Jupiter

Lying near Mars for the first half of the year, Jupiter is closest to the Earth on **10 February** and at opposition on **11 February**. The giant planet is in Leo until March; it strays into Cancer for April and then returns to Leo until it sinks into the twilight glow in July. In September, Jupiter reappears before dawn in Leo, and moves into Virgo during November.

Saturn

Saturn begins the year in Cetus and moves into Pisces in February, where it remains for the rest of the year. You can catch the ringworld to the west after sunset through to March. After disappearing behind the Sun, it reappears in the morning sky in June. Saturn is at opposition and at its closest to the Earth on **18 October**.

Uranus

The seventh planet lies in Taurus all year. Up until April, Uranus is visible in the evening sky: it re-emerges in the morning sky in July. Uranus reaches opposition and is closest to Earth on **30 November**.

Neptune

Neptune can be seen (though only through binoculars or a telescope) in Pisces in January and February, and then from June until the end of the year. The most distant planet is at its closest to the Earth on **27 September** and at opposition on **28 September**.

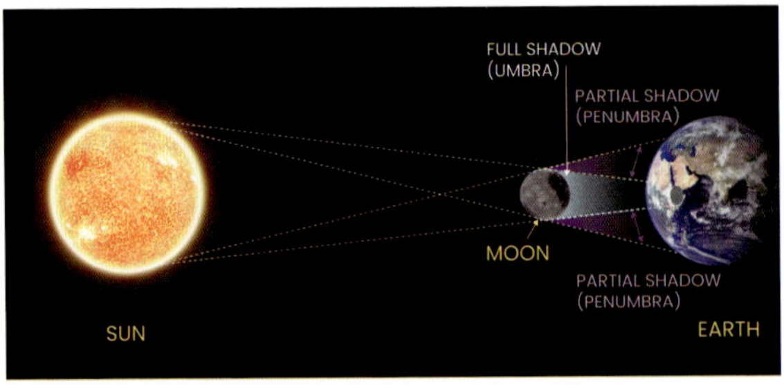

Where the dark central part (the umbra) of the Moon's shadow reaches the Earth, we are treated to a total solar eclipse. If the shadow doesn't quite reach the ground, we see an annular eclipse. People located within the penumbra observe a partial eclipse.

SOLAR ECLIPSES

On **6 February** an annular eclipse of the Sun is visible from parts of South America and western Africa, but nothing can be seen from Britain or Ireland.

A total solar eclipse, on **2 August**, will amaze viewers along a narrow band stretching through northern Africa and Arabia. The UK and Ireland will experience a partial solar eclipse, with 45 per cent of the Sun obscured in southern regions like Cornwall, falling to 20 per cent for the north of Scotland.

LUNAR ECLIPSES

There are no partial or total eclipses of the Moon in 2027.

METEOR SHOWERS

Shooting stars, or *meteors,* are tiny specks of interplanetary dust burning up in the Earth's atmosphere. At certain times of year, Earth passes through a stream of debris (usually left by a comet) and we see a *meteor shower*. The meteors appear to emanate from a point in the sky known as the *radiant*. Most showers are known by the constellation in which the radiant lies.

It's fun and rewarding to hold a meteor party. Get a group of friends together, each looking in a different direction: everyone calls out 'meteor' when they spot a shooting star. Record the time and brightness of each meteor, along with its path through the sky, And, of course, make a wish!

Table of annual meteor showers

Meteor shower	Date of maximum
Quadrantids	3/4 January
Lyrids	22/23 April
Eta Aquarids	7 May (am)
Perseids	12/13 August
Orionids	21/22 October
Leonids	17/18 November
Geminids	13/14 and 14/15 December

COMETS

Comets are dirty snowballs from the outer Solar System. If they fall towards the Sun, its heat evaporates their ices to produce a gaseous head (*coma*) and sometimes dramatic tails. Although some comets are visible to the naked eye, use binoculars to reveal stunning details in the coma and the tail.

Hundreds of comets move round the Sun in small orbits. But many more don't return for thousands or even millions of years. Most comets are now discovered in professional surveys of the sky, but a few are still found by dedicated amateur astronomers. No naked-eye comets are expected in 2027 at the time of writing, but we will witness the 66th appearance of Encke's Comet, visible in binoculars at the beginning of the year (see January's Topic and Special Events).

Here are some of the most popular sights in the night sky, in a season-by-season summary. It doesn't matter if you're a complete beginner, finding your way around the heavens with the unaided eye 👁 or binoculars; or if you're a seasoned stargazer, with a moderate telescope.
There's something here for everyone.

Each sky sight comes with a brief description, and a guide as to how you can best see it. Many of the most delectable objects are faint, so avoid moonlight when you go out spotting. Most of all, enjoy!

SPRING

Praesepe

Constellation: Cancer
Star Chart/Key: March; 5
Type/Distance: Star cluster; 600 light years
Magnitude: +3.7
A fuzzy patch to the unaided eye; a telescope reveals many of its 1000 stars.

M81 and M82

Constellation: Ursa Major
Star Chart/Key: March; 6
Type/Distance: Galaxies; 12 million light years
Magnitude: +6.9 (M81); +8.4 (M82)
A pair of interacting galaxies: the spiral M81 appears as an oval blur, and the starburst M82 as a streak of light.

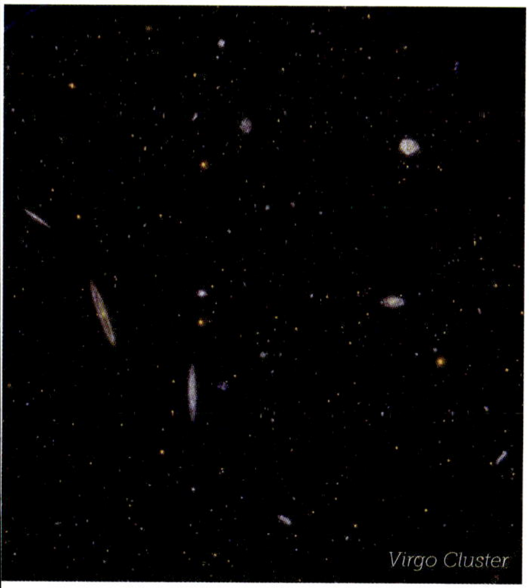

Virgo Cluster

The Plough

Constellation: Ursa Major
Star Chart/Key: April; 7
Type/Distance: Asterism; 82–123 light years
Magnitude: Stars are roughly magnitude +2
The seven brightest stars of the Great Bear form a large saucepan shape, called 'the Plough'.

Mizar and Alcor

Constellation: Ursa Major
Star Chart/Key: April; 8
Type/Distance: Double star; 83 & 82 light years
Magnitude: +2.0 (Mizar); +4.0 (Alcor)
The sky's classic double star, easily separated by the unaided eye: a telescope reveals Mizar itself is a close double.

Virgo Cluster

(difficult)

Constellation: Virgo
Star Chart/Key: May; 9
Type/Distance: Galaxy cluster; 54 million light years
Magnitude: Galaxies range from magnitude +9.4 downwards
Huge cluster of 2000 galaxies, best seen through moderate to large telescopes.

SUMMER

Antares

Constellation: Scorpius
Star Chart/Key: June; 10
Type/Distance: Red giant; 550 light years
Magnitude: +0.96
Bright red star close to the horizon. You can spot a faint green companion with a telescope.

M13

Constellation: Hercules
Star Chart/Key: June; 11
Type/Distance: Star cluster; 23,000 light years
Magnitude: +5.8
A faint blur to the naked eye, this ancient globular cluster is a delight seen through binoculars or a telescope. It boasts around half a million stars.

Lagoon and Trifid Nebulae

Constellation: Sagittarius
Star Chart/Key: July; 12
Type/Distance: Nebulae; 5000 light years
Magnitude: +6.0 (Lagoon); +7.0 (Trifid)
While the Lagoon Nebula is just visible to the unaided eye, you'll need binoculars or a telescope to spot the Trifid. The two are in the same binocular field of view, and present a stunning photo opportunity.

Albireo

Constellation: Cygnus
Star Chart/Key: August; 13
Type/Distance: Double star; 360 light years
Magnitude: +3.2 (Albireo A); +5.1 (Albireo B)
Good binoculars reveal Albireo as being double. But you'll need a small telescope to appreciate its full glory. The brighter star appears golden; its companion shines piercing sapphire. It is the most beautiful double star in the sky.

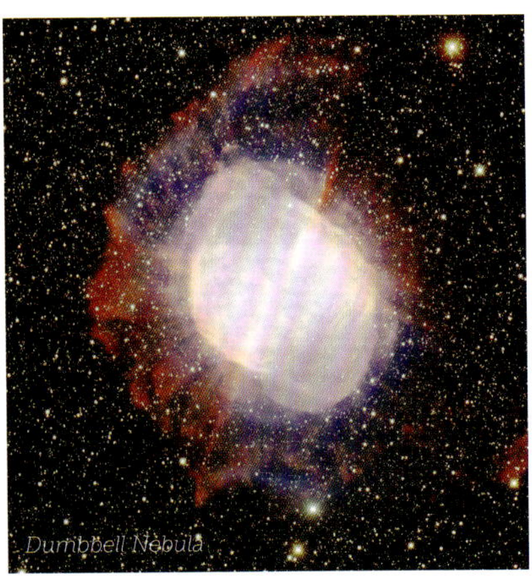
Dumbbell Nebula

Dumbbell Nebula

Constellation: Vulpecula
Star Chart/Key: August; 14
Type/Distance: Planetary nebula; 1360 light years
Magnitude: +7.5
Visible through binoculars, and a lovely sight through a small/medium telescope, this dying star has puffed off its atmosphere into space.

AUTUMN

Delta Cephei

Constellation: Cepheus
Star Chart/Key: September; 15
Type/Distance: Variable star; 890 light years
Magnitude: +3.5 to +4.4, varying over 5 days 9 hours
The classic variable star, Delta Cephei is chief of the Cepheids – stars that allow us to measure distances in the Universe (their variability time is coupled to their intrinsic luminosity). Visible to the unaided eye, but you'll need binoculars for serious observations.

Andromeda Galaxy

Constellation: Andromeda
Star Chart/Key: October; 16
Type/Distance: Galaxy; 2.5 million light years
Magnitude: +3.4
The nearest major galaxy to our own, the Andromeda Galaxy is easily visible to the unaided eye in unpolluted skies. Four times the width of the Full Moon, it's a great telescopic object and photographic target.

Mira

Constellation: Cetus
Star Chart/Key: November; 17
Type/Distance: Variable star; 300 light years
Magnitude: +2 to +10 over 332 days, although maxima and minima may vary.
Nicknamed 'the Wonderful', this distended red giant star is alarmingly variable as it swells and shrinks. At its brightest, Mira is a naked-eye object; binoculars may catch it at minimum; but you need a telescope to monitor this star. Its behaviour is unpredictable, and it's important to keep logging it.

Double Cluster

Constellation: Perseus
Star Chart/Key: November; 18
Type/Distance: Star clusters; 7500 light years
Magnitude: +3.7 and +3.8
A lovely sight to the unaided eye, these stunning young star clusters are sensational through binoculars or a small telescope. They're a great photographic target.

Algol

Constellation: Perseus
Star Chart/Key: November; 19
Type/Distance: Variable star; 90 light years
Magnitude: +2.1 to +3.4 over 2 days 21 hours
Like Mira, Algol is a variable star, but not an intrinsic one. It's an 'eclipsing binary' – its brightness falls when a fainter companion star periodically passes in front of the main star. It's easily monitored by the eye, binoculars or a telescope.

WINTER

Pleiades

Constellation: Taurus
Star Chart/Key: December; 20
Type/Distance: Star cluster; 444 light years
Magnitude: Stars range from magnitude +2.9 downwards
To the naked eye, most people can see six stars in the cluster, but it can rise to 14 for the keen-sighted. In binoculars or a telescope, they are a must-see. Astronomers have observed 1000 stars in the Pleiades.

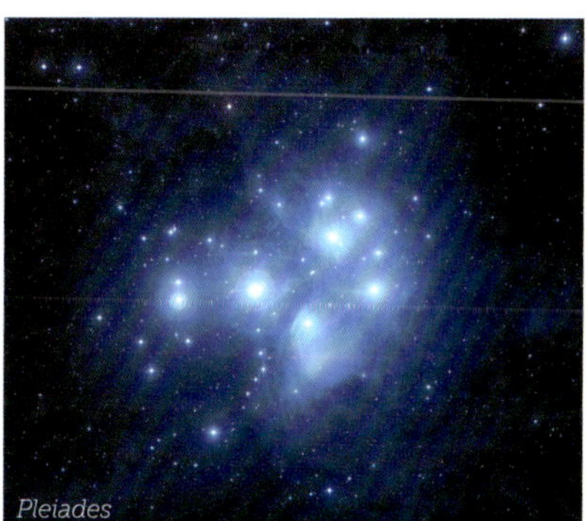

Pleiades

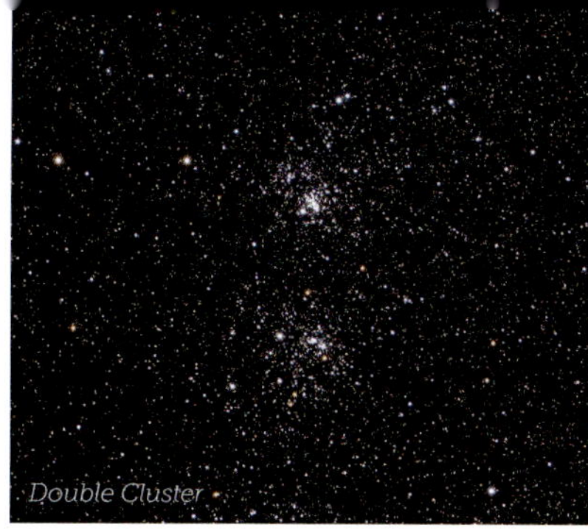

Double Cluster

Orion Nebula

Constellation: Orion
Star Chart/Key: January; 1
Type/Distance: Nebula; 1340 light years
Magnitude: +4.0
A striking sight even to the unaided eye, the Orion Nebula – a star-forming region 24 light years across – hangs just below Orion's Belt. Through binoculars or a small telescope, it is staggering. A photographic must!

Betelgeuse

Constellation: Orion
Star Chart/Key: January; 2
Type/Distance: Variable star; 720 light years
Magnitude: 0.0 to +1.6
Even with the unaided eye, you can see that Betelgeuse is slightly variable over months, as the red giant star billows in and out.

M35

Constellation: Gemini
Star Chart/Key: February; 3
Type/Distance: Star cluster; 2800 light years
Magnitude: +5.3
Just visible to the unaided eye, this cluster of around 2000 stars is a lovely sight through a small telescope.

Sirius

Constellation: Canis Major
Star Chart/Key: February; 4
Type/Distance: Double star; 8.6 light years
Magnitude: –1.5
You can't miss the Dog Star. It's the brightest star in the sky! But you'll need a 150-mm reflecting telescope (preferably bigger) to pick out its +8.4 magnitude companion – a white dwarf nicknamed 'the Pup'.

OBSERVING THE SUN

BY ROBIN SCAGELL

There is one astronomical object that is potentially dangerous to observe – the Sun. Its light and heat can seriously damage both eyesight and telescopes, so observers need to be particularly careful. Fortunately, these days there are plenty of options for observing our nearest star safely, from budget to megabuck.

What can you see on the Sun? It is more than just a bland disc of light. Easiest to observe are sunspots, which are darker areas on the Sun. They are never the same for two days running, as they are carried around the disc by the Sun's rotation. During 2027 there are likely to be spots visible on most days as the Sun remains quite active, but predictions are that the numbers of sunspots will decline over the next few years and then rise again.

One means of safely observing the Sun recommended in the past was to project the Sun's image through a telescope onto a piece of card held some distance behind the eyepiece. However, this results in focusing all the Sun's heat inside the eyepiece, which could damage its internal components. There is also a risk that someone may inadvertently look into the telescope, so these days the aim is to reduce the heat from the Sun before it enters the telescope.

To see sunspots safely, you need to cut down the brilliance of the Sun by a factor of about 100,000. But it's crucial to do this using the proper type of filter. Some domestic materials that look as if they might work, such a black bin liner, cut down the white sunlight but allow through infrared heat radiation, which will literally burn your eyes. To make matters worse, your natural reaction to avoid a bright light is no longer present, so before you know what's happening, it's too late.

Fortunately, solar filters are widely available from telescope suppliers in a variety of sizes to fit over the front of telescopes, converting an everyday telescope into what's referred to as a white-light solar instrument – that is, it shows the Sun in its natural colour as seen by eye, which is essentially white. The filter material itself, such as Baader AstroSolar® Safety Film, is also available in A4 sheets of the thin film material, but if you decide to make your own filter, make absolutely sure that it covers every part of the telescope's aperture and is completely secure. It can only safely be used over the top end of the telescope. Don't be tempted to use it at the eyepiece end of the telescope, as the telescope lens or mirror acts like a magnifying glass and will very quickly burn a hole in the thin film.

You can also buy ready-made solar telescopes and binoculars for observing sunspots. These have the dense filter pre-installed within the instrument, so there is no risk of it becoming dislodged, and offer the safest way of observing the Sun and its spots. One such instrument is the Acuter Solarus 80, an 80-millimetre telescope. Like all solar telescopes it is available as the optical tube assembly (OTA) only, for those who already have a suitable mounting, or on a tripod with a mounting that can be driven to track the Sun.

As with any telescope, the mounting is just as important as the instrument

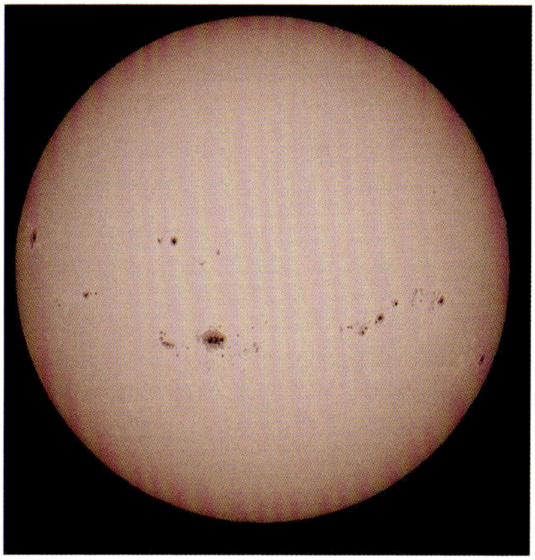

A view of the Sun in white light in August 2024, when the Sun was particularly active, taken through a 60-mm refractor with a solar filter covering the front lens. John Chapman-Smith of Skegness used a Canon 600D camera for a single 1/80 second exposure at ISO 200.

itself. If you are about to buy your first telescope, aim to get one that fixes to the mount using a standard dovetail, as this will allow you to swap instruments very easily. For solar observing, what is called an altazimuth mount is all you need. This has basic up-and-down and side-to-side movements rather than the more complex equatorial mounts which are more suited to long-exposure astrophotography. But don't be tempted to economise by fitting the telescope to a standard photographic tripod instead. These rarely provide the stability or the ease of movement needed.

With a white-light instrument you can track the movement of sunspots across the solar disc, taking over a week to do so, and view what are called faculae, which are brighter areas often seen near the edge, or limb, of the Sun's disc. The bright disc itself is fairly uniform though slightly darker at the limb, but on closer inspection you might see the fine granulation on the Sun which is caused by convection cells rising and falling.

The Acuter Solarus 80 comes with an adapter that allows you to point a mobile phone into the eyepiece to take a snap. Its zoom eyepiece is not interchangeable, however, unlike most astronomical telescopes, so it isn't possible to fit a more advanced astro-camera.

If you just want to take images of the solar disc in white light, consider the option of a smart telescope, such as a Dwarf 3 or Seestar. These popular instruments are controlled using an app on a phone or tablet, and as well as taking excellent photos of deep-sky objects they usually come with a solar filter that will take white-light pictures of the Sun of medium quality completely safely.

Viewing the Sun in white light, and keeping track of the ebb and flow of sunspots, is fascinating. But you might expect to also see the Sun with what look

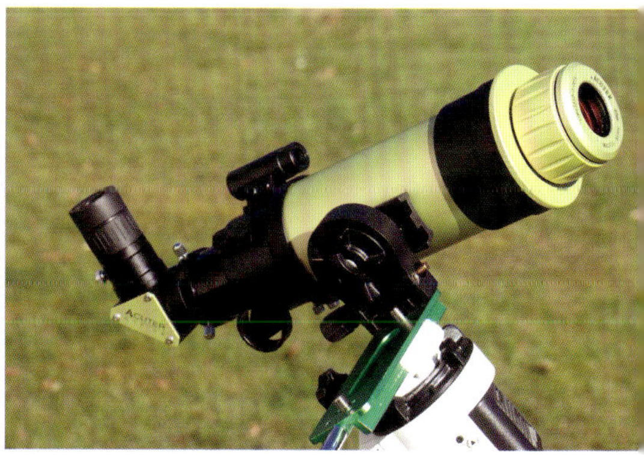

The etalon on this Acuter Elite Phoenix 80 is 40 mm in diameter and is mounted on the front of an 80-mm refracting telescope. It can be unscrewed easily to convert the instrument to a conventional telescope. In this case it is mounted on a Sky-Watcher Star Adventurer equatorial tracking mount that may also be used for basic deep-sky imaging.

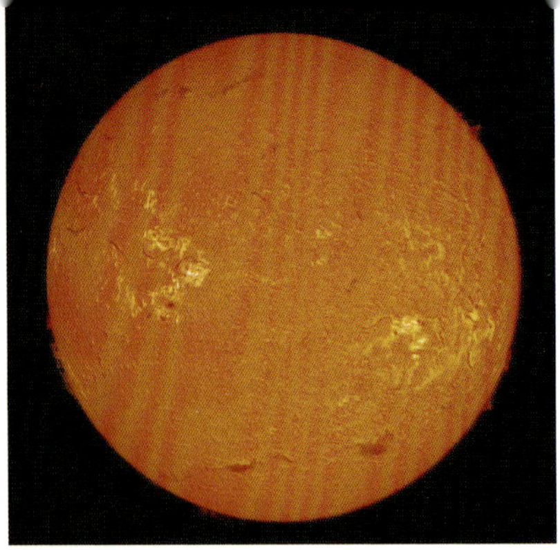

Kwong Man of Edgware in the London suburbs used his Acuter Elite Phoenix 80 to take this H-alpha solar image on 30 September 2025. Sunspots and filaments are well defined, and prominences are visible around the limb. He used a mono astro-camera and added the colour in processing.

like giant flames around its edges, as are invariably shown in artworks of the Sun and in time-lapse movies. People often think these are called flares, but they are actually prominences. They look spectacular, but compared with the brilliance of the Sun they are quite faint as their light comes only from hydrogen atoms in a very narrow band of the spectrum, in deep red light, at a wavelength referred to as hydrogen alpha, or H alpha.

Until the 20th century, the only occasions on which prominences could be seen were during total solar eclipses, when the light from the Sun's disc is hidden by the Moon. Eventually it became possible to view them at any time using optical devices called etalons, which transmit a narrow band of H-alpha light, plus other unwanted parts of the spectrum. These other parts need to be removed using a separate filter known as a blocking filter. Etalons have become much more affordable in recent years, and now there is a wide choice of instruments, most of which are purpose-built H-alpha solar scopes.

Seen through an H-alpha telescope, the Sun is the deep orange-red colour of this wavelength of light, which lies right at the edge of human colour vision. At first glance you see a noticeably mottled solar disc, possibly with snake-like features crossing it. Around the limb it is likely that you will see a few prominences, of varying sizes. They may not be very striking to begin with, but it is possible to fine-tune the wavelength transmitted by the etalon to make them more obvious, simply by turning an adjuster. Some prominences are huge, with lengths similar to the diameter of the Sun itself, but even the smaller and more common ones can have interesting and dramatic shapes. One type is referred to as a hedgerow prominence, on the basis of its appearance. Prominences change in shape only slowly, over a period of many minutes or hours. Solar details and prominences are visible in hydrogen alpha even when the Sun's activity is low and sunspots are small or even absent.

Key features of dedicated solar scopes are the aperture, which generally ranges from 40 millimetres up to 80 millimetres, the bandwidth of the etalon and the evenness and size of the field of view. In some instruments at the cheaper end of

The Lunt LS50THa B600PT is a 50-mm H-alpha instrument with a 6-mm blocking filter suitable for both visual and imaging use. It can be converted to a double-stacked version if required.

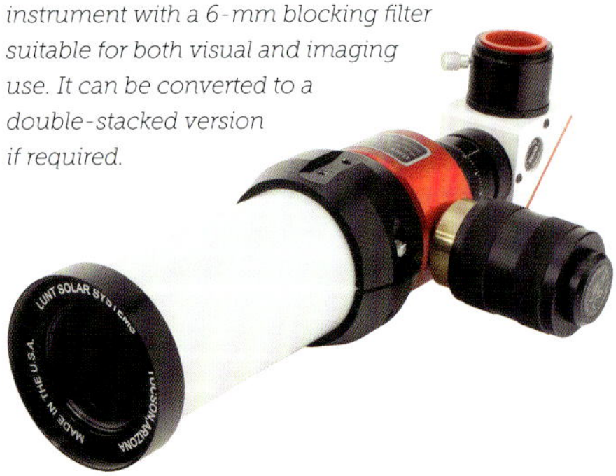

the range you may need to adjust the tuning to change the position of a 'sweet spot' of optimum performance across different parts of the solar disc, or the disc may not be evenly illuminated. Etalons considerably increase in cost with size, so many solar scopes use etalons inside the tube, where the beam of light is narrower, rather than over the main aperture as with a white-light filter. The diameter of the blocking filter affects the size of the field of view.

The narrower the bandwidth of light transmitted by the etalon, the better the view of the features. The bandwidth is usually described in the otherwise outmoded unit of angstroms (abbreviated Å), rather than the normal unit of nanometres (abbreviated nm). An angstrom is a tenth of a nanometre. A single etalon will transmit a bandwith of 0.7 Å, which gives an impressive view. But many devotees of solar viewing prefer to double-stack etalons, which reduces the brightness of the view but cuts the bandwidth to around 0.5 Å, giving a sharper and more contrasty view.

Hydrogen-alpha solar scopes are all expensive compared with their conventional counterparts, starting at prices similar to those of full-featured 85-inch TVs with the larger double-stacked versions going into several thousands of pounds. And of course there is only one object that they can be used to observe – the Sun. It is possible, however, to adapt a conventional telescope by using an etalon assembly at the focus position in place of the eyepiece. The DayStar Quark will convert a standard refracting telescope into a solar telescope, and includes the blocking filter, although additional filters may be needed for larger apertures. It is available in two versions, optimised for viewing either the Sun's disc or the prominences.

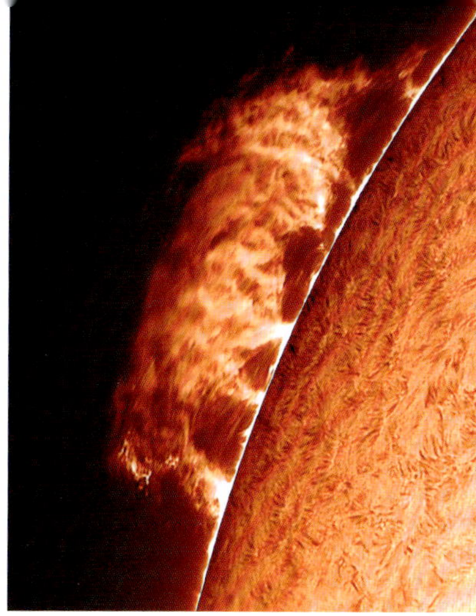

Carl Bowron of Doncaster uses a 120-mm Sky-Watcher refracting telescope and DayStar Quark Chromosphere filter to take high-resolution images of small areas of the Sun. He photographed this large prominence on 11 July 2025.

Another option is the Acuter Elite Phoenix Solar Telescope, which uses a 40-millimetre aperture etalon attached to the front of a conventional 80-millimetre refracting telescope. The etalon can be unscrewed so that the telescope can be used as a standard astronomical telescope. For solar viewing it must be used with a star diagonal containing a blocking filter, to provide a convenient viewing position, but for night use this can be replaced by a standard non-filtered star diagonal to give good views of deep-sky objects. The instrument is provided with a finder device that projects a pinhole image of the Sun onto a bullseye-type target so that it can be pointed at the Sun without requiring the user to squint along the tube. As with dedicated H-alpha scopes, it can also be used with an astro-camera for solar imaging, or for deep-sky photography using the appropriate techniques.

ASTRO PLACES & EVENTS

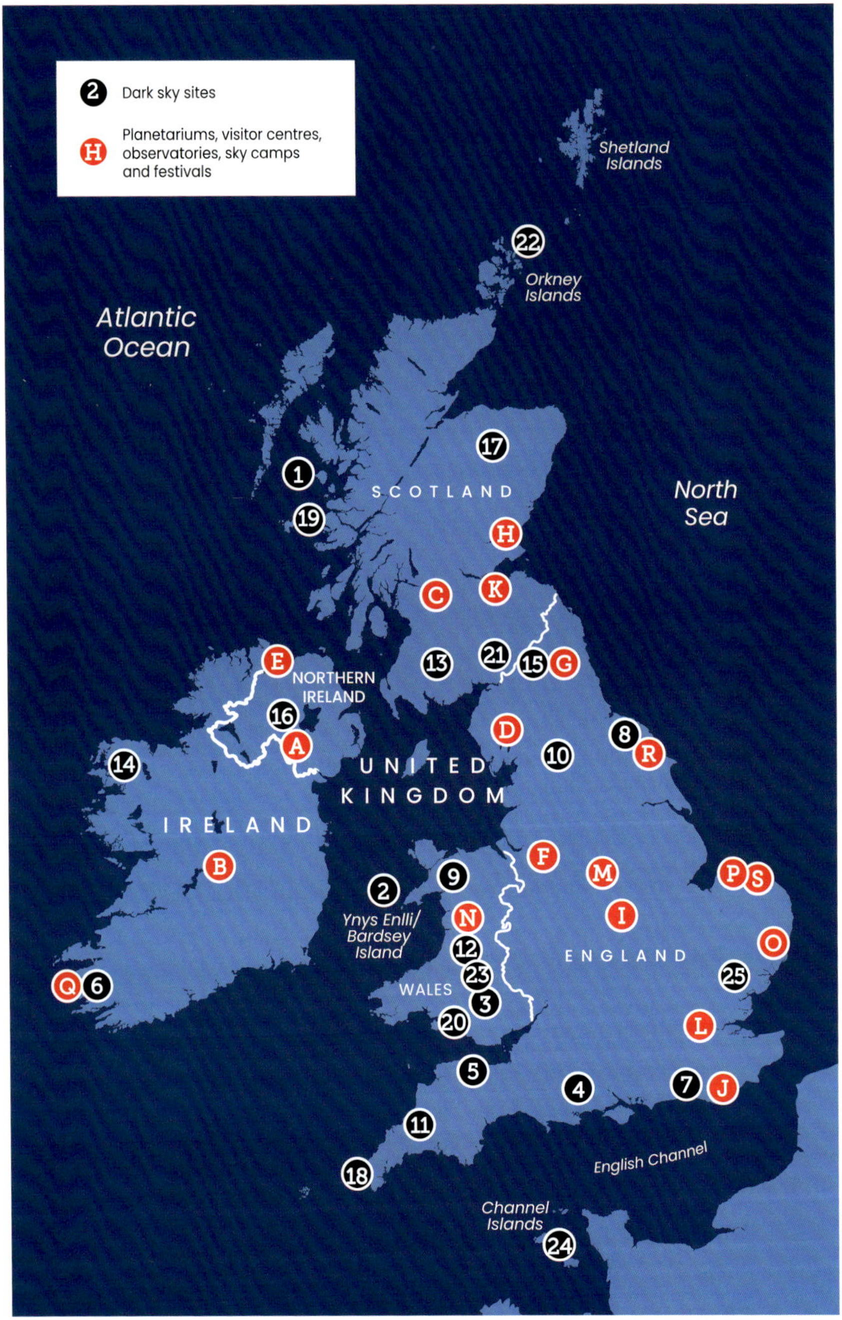

Take a break from your backyard stargazing! View the pristine heavens from a protected dark-sky site, visit a major observatory or planetarium, or join in the fun at a star camp or astronomical event.

DARK-SKY SITES

The International Dark-Sky Association, based in Tucson, Arizona, checks out the darkest places in the world. In these islands, 25 places are internationally recognised for their unsullied view of the heavens or their commitment to combating light pollution. Many are in National Landscapes, formerly Areas of Outstanding Natural Beauty.

DARK SKY SANCTUARIES

Noted for exceptionally dark nights and a nocturnal environment protected for its scientific and cultural heritage, a Dark Sky Sanctuary is often in a remote location.

1. Isle of Rum

Designated: 2024 • *Area:* 105 sq km
Nearest town: Mallaig
Accessible only to walkers and cyclists, the Isle of Rum is home to 120,000 pairs of Manx shearwaters as well as golden eagles and white-tailed sea eagles. It has only 40 inhabitants and no public lighting.

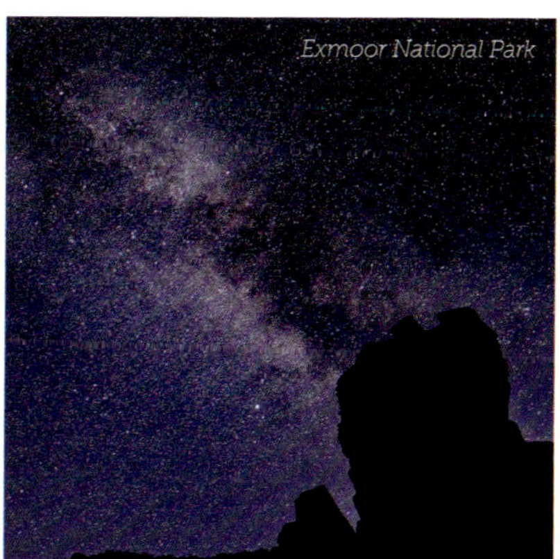

Exmoor National Park

2. Ynys Enlli/Bardsey Island

Designated: 2023 • *Area:* 1.8 sq km
Nearest town: Pwllheli
Europe's first designated Dark Sky Sanctuary is home to a major bird observatory, with an emphasis on nocturnal species. It lies 3 km off the Welsh coast, and is accessible only by small boat during good weather.

DARK SKY RESERVES

Dark Sky Reserves are generally large, and consist of a very dark core region surrounded by a peripheral area where lighting is minimal. Eight out of a total of 19 worldwide are in Britain and Ireland.

3. Bannau Brycheiniog/ Brecon Beacons National Park

Designated: 2013 • *Area:* 1347 sq km
Nearest towns: Brecon, Merthyr Tydfil
Situated in the mountains of South Wales, where sheep outnumber humans 30 to 1, this Reserve is home to 33,000 people – yet lighting is controlled so that the core zone has some of the darkest skies in Wales.

4. Cranborne Chase

Designated: 2019 • *Area:* 981 sq km
Nearest towns: Salisbury, Shaftesbury
Not far from Stonehenge, Cranborne Chase is a chalk plateau rising to 277 m at Win Green. This National Landscape has commanding views to the west and to some extent to the north.

5. Exmoor National Park

Designated: 2011 • *Area:* 693 sq km
Nearest towns: Barnstaple, Minehead, Taunton
The first Dark Sky Reserve site to be designated in these islands, Exmoor is a moorland area with much preserved history and many monuments within the 81-sq-km core zone.

6. Kerry

Designated: 2014 • *Area:* 700 sq km
Nearest towns: Kenmare, Waterville
Its location between the Kerry Mountains and the Atlantic Ocean gives this Dark Sky Reserve natural protection from light pollution, yet it's readily accessible from the Wild Atlantic Way, the stunning tourist route that runs through the reserve.

7. Moore's Reserve (South Downs)

Designated: 2016 • *Area:* 1627 sq km
Nearest towns: Brighton, Portsmouth
Named after the British astronomy populariser Sir Patrick Moore (1923–2012) who lived nearby, this reserve is sandwiched between London and the busy seaside resorts of Brighton and Worthing – yet it retains remarkably dark skies.

8. North York Moors National Park

Designated: 2020 • *Area:* 1440 sq km
Nearest towns: Scarborough, Whitby
Despite its proximity to the busy tourist destinations of Whitby and Scarborough, the North York Moors is a largely deserted expanse of heather and bog moorland, with wide views of the night sky.

9. Eryri/Snowdonia National Park

Designated: 2015 • *Area:* 2132 sq km
Nearest towns: Harlech, Porthmadog

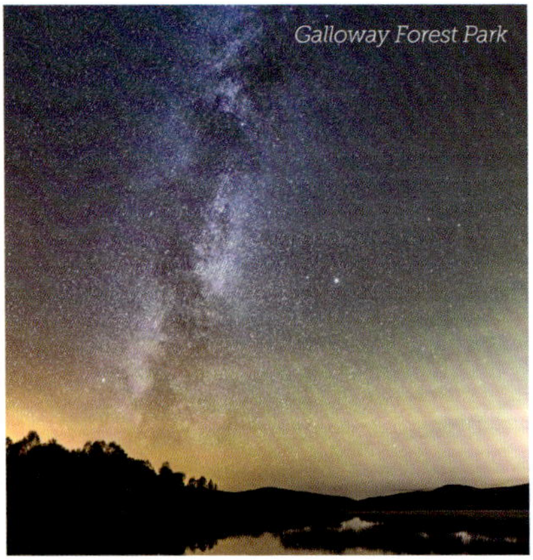

Galloway Forest Park

The national park encompasses around 10 per cent of the total land area of Wales, and the darkest skies are to be seen from the rugged central area around Mount Snowdon/ Yr Wyddfa (1085 m).

10. Yorkshire Dales National Park

Designated: 2020 • *Area:* 2180 sq km
Nearest towns: Hawes, Kirby Lonsdale
The largest dark-sky site in Britain and Ireland, the Yorkshire Dales National Park boasts impressive waterfalls and caves, and brilliant night skies within reach of major cities like Leeds and Manchester.

DARK SKY PARKS

Smaller regions with exceptionally low light pollution are designated Dark Sky Parks. There are currently eight in Britain and Ireland.

11. Bodmin Moor Dark Sky Landscape

Designated: 2017 • *Area:* 208 sq km
Nearest towns: Bodmin, Launceston, Liskeard
This remote, rugged area of granite moorland in north-east Cornwall is a working agricultural landscape, protecting it from large-scale development that would threaten dark skies.

12. Elan Valley Estate

Designated: 2015 • *Area:* 180 sq km
Nearest towns: Aberystwyth, Rhayader
The city of Birmingham purchased this Welsh valley in 1892 to construct reservoirs that would provide a regular supply of pure water. Views of the starry night over the reservoirs are particularly impressive.

13. Galloway Forest Park

Designated: 2009 • *Area:* 780 sq km
Nearest towns: Girvan, Newton Stewart
Galloway Forest Park is the largest forest park in the UK, and 20 per cent has been set aside as a core area with no illumination allowed. It's a mecca not just for astronomers but for nocturnal wildlife whose lives are often disrupted elsewhere by light pollution.

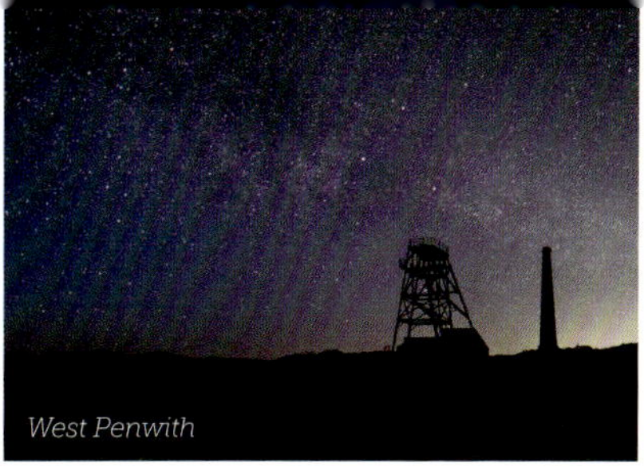

West Penwith

14. Mayo Dark Sky Park

Designated: 2016 • *Area:* 150 sq km
Nearest towns: Ballina, Castlebar
One of the largest expanses of peat landscape in Europe, this region supports a unique diversity of bog-dwelling species. Unsuitable for agriculture, and adjoining the Atlantic Ocean, the park should enjoy pristine skies far into the future.

15. Northumberland National Park and Kielder Water & Forest Park

Designated: 2013 • *Area:* 1592 sq km
Nearest towns: Jedburgh, Rothbury
Near Hadrian's Wall, built to keep the Picts from Roman Britain, this Dark Sky Park was designated as a bulwark against light pollution invading the darkness of northern England. It contains the largest reservoir and most extensive forest in northern Europe.

16. OM Dark Sky Park & Observatory

Designated: 2020 • *Area:* 15 sq km
Nearest towns: Cookstown, Magherafelt
The first dark-sky place accredited in Northern Ireland, OM is set among rolling hills and is centred on the Bronze Age site of Beaghmore Stone Circles.

17. Tomintoul and Glenlivet, Cairngorms

Designated: 2018 • *Area:* 230 sq km
Nearest towns: Dufftown, Grantown-on-Spey
Containing the dramatic landscape of the Cairngorm Mountains, this Dark Sky Park is home to Britain's only herd of wild reindeer. And, if the weather is cloudy, the Park also contains the Glenlivet whisky distillery!

18. West Penwith

Designated: 2021 • *Area:* 136 sq km
Nearest towns: Penzance, St Ives
The very western tip of Cornwall, stretching down to Land's End, West Penwith is a wild landscape, with stunning sea views and ancient archaeological remains – some thought to be astronomically aligned.

DARK SKY COMMUNITIES

A town, village or complete island that's actively fighting light pollution can be designated a Dark Sky Community, with six listed in Britain and Ireland so far, plus one in the Channel Islands.

19. Coll

Designated: 2013 • *Area:* 77 sq km
The Scottish island of Coll is home to just 200 permanent residents, plus myriad birds in its extensive nature reserve. The island has adopted a light-management plan to ensure its skies remain dark.

20. Gower National Landscape

Designated: 2025 • *Area:* 186 sq km
A wild peninsula near Swansea, Gower is known for its stunning landscapes. Now, all its 1641 streetlights have been replaced with dark-sky-approved lamps to minimise light pollution.

21. Moffat

Designated: 2016 • *Area:* 147 sq km
This former spa town is a tourist base for southern Scotland. It has strict outdoor lighting policies to reduce light pollution in its hinterland.

22. North Ronaldsay

Designated: 2021 • *Area:* 7 sq km
The northernmost island in Orkney. Many visitors come to its bird observatory, and the island's appeal extends to astronomers with its Dark Sky Community designation.

23. Presteigne & Norton Dark Sky Community

Designated: 2023 • *Area:* 40 sq km
Two neighbouring towns in Wales combined to severely curtail night-time lighting in an area that's home to the Spaceguard Centre observatory and to colonies of light-sensitive bats.

24. Sark

Designated: 2011 • *Area:* 5 sq km
One of the Channel Islands, Sark was Europe's first Dark Sky Community. Its pitch-black skies are largely due to the island's ban on public lighting and motor vehicles apart from tractors.

25. Thorington Street

Designated: 2026 • *Area:* 3.5 sq km
The hamlet of Thorington Street in Dedham Vale (featured in John Constable's classic paintings) became England's first Dark Sky Community after a campaign involving residents, schools and businesses.

PLANETARIUMS, VISITOR CENTRES AND OBSERVATORIES

For a great day out under the cloudless sky of a planetarium, a tour of a world-beating observatory or an evening observing the stars, here are some leading locations. Many venues combine a planetarium (), visitor centre (V) and observatory ().

For smaller observatories near to you, check out the comprehensive list at gostargazing.co.uk.

A. Armagh Observatory and Planetarium

The longest-running planetarium and the second-oldest observatory in the UK. Combine a digital planetarium show with a tour of the observatory's historical instruments.

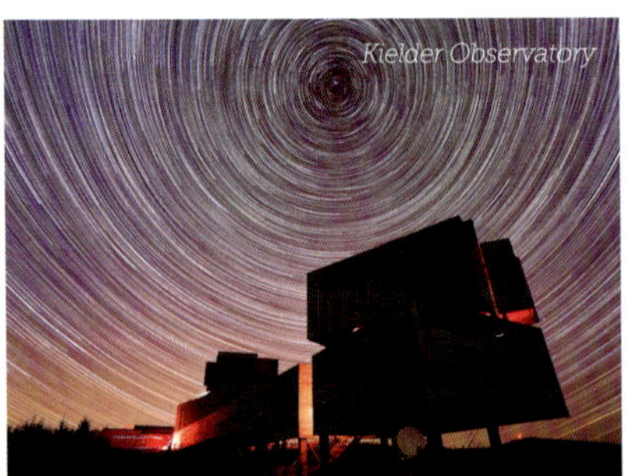

Kielder Observatory

B. Birr Castle Demesne V

Marvel at the well-preserved remains of the Leviathan of Parsonstown, the largest telescope in the world from 1845 to 1917. The Science Centre chronicles how the Third Earl of Rosse created his great instrument, and discovered the spiral shape of galaxies.

C. Glasgow Science Centre Planetarium

One of Scotland's most popular visitor attractions, Glasgow Science Centre features a 15-m diameter planetarium. After viewing the night sky in comfort, you can be awed by the giant screen of the Centre's IMAX cinema.

D. Grizedale Observatory

Right in the heart of the Lake District, the observatory has a fleet of 200-mm Dobsonian telescopes, two 400-mm reflectors and a robotic telescope allowing you to view the Universe from indoors, along with a state-of-the-art planetarium.

E. Inishowen Planetarium

Part of the Inishowen Maritime Museum, the planetarium is picturesquely set beside Lough Foyle. As well as putting on astronomy shows, the Maritime Museum uses the planetarium dome to project immersive ocean experiences.

F. Jodrell Bank Discovery Centre V

Grab a close-up view of the Lovell Telescope, the great radio dish that tracked rockets in the early days of the space race, downloaded pictures from the Moon and now investigates the secrets of pulsars. Other displays include exhibitions, talks and interactive activities.

G. Kielder Observatory

Sited under some of the darkest skies in England, Kielder has a variety of telescopes for visual observing and astrophotography. It hosts some 700 events per year.

H. Mills Observatory

On a hill above Dundee, the Mills Observatory was the UK's first purpose-built public observatory, and it carries on opening its doors to the public every clear weeknight. There's also a small planetarium and a gift shop.

I. National Space Centre

Located on the outskirts of Leicester, this visitor centre focuses on the history – and future – of the UK in space exploration. Explore the wider Universe in the 192-seat Sir Patrick Moore planetarium.

J. Observatory Science Centre

On a hillside at Herstmonceux in Sussex, the Observatory Science Centre is located within a cluster of green domes that once housed the telescopes of the Royal Greenwich Observatory. As well as fascinating exhibits, the observatory holds regular stargazing evenings.

K. Royal Observatory, Edinburgh

Scotland's premier observatory no longer makes professional observations: its astronomers now build and use large telescopes on Hawai'i and in Chile. But the Edinburgh site is still active, with a visitor centre, regular astronomical talks and public stargazing evenings.

L. Royal Observatory and Peter Harrison Planetarium

The home of British astronomy, the Royal Observatory at Greenwich is a fascinating museum of astronomy and timekeeping: stand on the Meridian Line, with a foot in each hemisphere! The planetarium hosts a variety of astronomical shows.

M. Sherwood Observatory

Set in Robin Hood country near Mansfield, Nottinghamshire, the observatory offers night-time observing sessions during autumn and winter, and solar observations in spring and summer. There's also a planetarium with a wide variety of shows.

STAR CAMPS AND FESTIVALS

Whether you're a beginner or a seasoned sky-watcher, enjoy a weekend under the stars with fellow astronomers at these annual star camps and festivals. Also, check out events at the countries' dark-sky sites (pages 91–94) as many of them host Dark Skies Festivals. And there's a comprehensive list of sky-watching activities at gostargazing.co.uk.

N. Cambrian Mountains Stargazing Weekends

Staylittle, Powys • Most months
Stay in a deluxe wigwam cabin under the dark skies of Mid Wales, for a weekend of stargazing, plus tuition in astronomy and astrophotography.

O. Haw Wood Farm Astronomers' Week

Saxmundham, Suffolk • October
Astronomers literally pitch up at Haw Wood for a small-scale, friendly week of stargazing, organised by local astronomical societies.

P. Kelling Heath Star Party

Kelling Heath, Norfolk • March and September
Claiming to be the largest star party in the UK, this biennial event fills three fields with astronomers, tents, trade stands and top-end amateur telescopes.

Q. Skellig Star Party

Ballinskelligs, County Kerry • August
Held by the shores of the Atlantic Ocean, Ireland's premier star party features talks by leading astronomers as well as night-time observing.

R. StarFest

Dalby Forest, North Yorkshire • August
This event attracts astronomers from all over the country for a weekend of events that can include rocket-building, talks, an astronomical pub quiz and – of course! – observing the sky.

S. WinterFest Astro Star Party

Kelling Heath, Norfolk • November
Organised by the Birmingham Astronomical Society around the time of New Moon, to give its members some respite from the city's lights, WinterFest is open to all who want to view the glittering winter stars.

THE AUTHOR

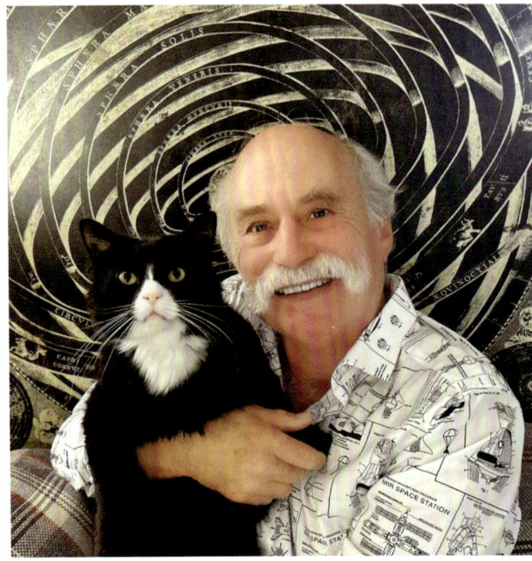

Professor **Nigel Henbest** is an award-winning British science writer, specialising in astronomy and space. After research in radio astronomy at Cambridge, he became a consultant to both the Royal Greenwich Observatory and *New Scientist* magazine.

Originally with Heather Couper, Nigel has been writing Philip's *Stargazing* since 2005. The author of 50 other books, Nigel also co-founded a TV production company where he produced acclaimed programmes and international series on astronomy and space.

Now an Adjunct Professor at the University of North Carolina, Wilmington, Nigel is a Fellow of the Royal Astronomical Society and a Future Astronaut with Virgin Galactic; asteroid 3795 is named 'Nigel' in his honour.

Married with two stepdaughters, Nigel lives in Buckinghamshire and North Carolina.

ACKNOWLEDGEMENTS

PHOTOGRAPHS

Alamy Stock Photo: Anthony P. Morris, Farmoor 30, **Chrispo** 91, **Gary Cook** 92, **Keith Mason** 93; **Carl F Bowron** 89; **Courtesy Bresser UK** 88b; **John Chapman-Smith** 87a; **Stephen Doody** 35; **Josh Dury Photo-Media** 17; **Getty Images: Thilina Kaluthotage/NurPhoto** 78; **Jonathen Harty** 71; **Tracey Harty** 52; **Bridget Henbest** 96; **Nigel Henbest** 7; **Kielder Observatory Astronomical Society/Dan Monk** 94; **K.W.Man** 88a; **Mary McIntyre** 77; **Bill McSorley, West Yorkshire Astronomical Society, Messier** 44, **The Beehive Cluster (Praesepe)** 23; **Abby Moule** 41; **NASA: Davide De Martin & the ESA/ESO/NASA Photoshop FITS Liberator** 85b, **ESA, J. Clarke (Boston Univ.) and G. Bacon (STScI)** 48, **Goddard** 49, **JPL 60, JPL-Caltech** 37, 55r, 79br, **JPL-Caltech/SSC** 83, **UCLA 2, University of Arizona** 43; **NOIRLab: T.A. Rector (University of Alaska Anchorage) and H. Schweiker (WIYN and NOIRLab/NSF/AURA)** 84; **Damian Peach** 47; **Martin Pugh** 1; **Robin Scagell** 6, 87b; **Sara Wager, swagastro.com** 29; **Tim White** 65; **Wikimedia Commons: Brocken Inaglory (CC by-SA 3.0)** 66, **David Moug** 42, **Genuson (CC by-SA 3.0)** 85a; **Tom Williams** 11, 59.

ARTWORKS

Star Charts: Wil Tirion/Philip's with extra annotation by Philip's.

Special Event Charts: Nigel Henbest/ Stellarium (www.stellarium.org).

Pages 80–82: Chris Bell.

Page 90: Philip's